Chromatography of Pharmaceuticals

ACS SYMPOSIUM SERIES **512**

Chromatography
of
Pharmaceuticals

Natural, Synthetic, and Recombinant Products

Satinder Ahuja, EDITOR
Ciba–Geigy Corporation

Developed from a symposium sponsored
by the Division of Analytical Chemistry
of the American Chemical Society
at the Fourth Chemical Congress of North America
(202nd National Meeting of the American Chemical Society),
New York, New York,
August 25–30, 1991

American Chemical Society, Washington, DC 1992

Seplae
Chem

Library of Congress Cataloging-in-Publication Data

Chromatography of pharmaceuticals: natural, synthetic, and recombinant products / Satinder Ahuja, editor.

    p.   cm.—(ACS symposium series, ISSN 0097–6156; 512)

"Developed from a symposium sponsored by the Division of Agricultural and Food Chemistry of the American Chemical Society at the Fourth Chemical Congress of North America (202nd National Meeting of the American Chemical Society), New York, New York, August 25–30, 1991."

Includes bibliographical references and index.

ISBN 0–8412–2498–6

1. Chromatography—Congresses.   2. Pharmaceutical   chemistry—Congresses. 3. Drugs—Analysis—Congresses.

I. Ahuja, Satinder, 1933–   . II. American Chemical Society. Division of Agricultural and Food Chemistry. III. American Chemical Society. Meeting (202nd: 1991: New York, N.Y.) IV. Chemical Congress of North America (4th: 1991: New York, N.Y.) V. Series.

[DNLM: 1. Chromatography—methods—congresses. 2. Drugs—analysis—congresses. GV 25 C5575 1991]

RS189.5.C48C53   1992
615.1901—dc20
DNLM/DLC
for Library of Congress   92–49380
                                                  CIP

The paper used in this publication meets the minimum requirements of American National Standard for Information Sciences—Permanence of Paper for Printed Library Materials, ANSI Z39.48–1984. ∞

Foreword

THE ACS SYMPOSIUM SERIES was first published in 1974 to provide a mechanism for publishing symposia quickly in book form. The purpose of this series is to publish comprehensive books developed from symposia, which are usually "snapshots in time" of the current research being done on a topic, plus some review material on the topic. For this reason, it is necessary that the papers be published as quickly as possible.

Before a symposium-based book is put under contract, the proposed table of contents is reviewed for appropriateness to the topic and for comprehensiveness of the collection. Some papers are excluded at this point, and others are added to round out the scope of the volume. In addition, a draft of each paper is peer-reviewed prior to final acceptance or rejection. This anonymous review process is supervised by the organizer(s) of the symposium, who become the editor(s) of the book. The authors then revise their papers according the the recommendations of both the reviewers and the editors, prepare camera-ready copy, and submit the final papers to the editors, who check that all necessary revisions have been made.

As a rule, only original research papers and original review papers are included in the volumes. Verbatim reproductions of previously published papers are not accepted.

M. Joan Comstock
Series Editor

Contents

INDEXES

Preface

PHARMACEUTICAL ANALYSES ARE PERFORMED in a variety of situations that range from the assay of a new chemical entity in the presence of optical isomers and other related compounds (e.g., intermediates, byproducts, degradation products, and metabolites) to complex determination at trace or ultratrace concentrations in various matrices. Chromatography plays a major role in pharmaceutical analyses because it provides selectivity and detectability. *Chromatography of Pharmaceuticals*, which is intended to serve as a reference for pharmaceutical researchers, surveys the use of chromatography in the analysis of pharmaceuticals.

A number of analytical methods are discussed in great detail. These include gas chromatography (GC), capillary GC, GC–mass spectrometry (MS), high-performance liquid chromatography (HPLC or LC), LC–LC–ultraviolet (UV) detection, LC–MS, LC–MS–MS, capillary electrophoresis (CE), robotics, and preparative separations. Alternative strategies for analyzing pharmaceutical compounds are also discussed, along with the popular reversed-phase HPLC methods.

The purity of recombinant proteins is of paramount importance in drug therapy. Impurities in these drugs can be potentially unsafe and can include endotoxins, DNA, and host-cell proteins. The ability to detect, identify, and quantify impurities is an ongoing challenge for the analyst. This book provides an overview of the current chromatographic methods used to analyze host-cell protein impurities.

Biochemical interactions, such as those of binding proteins with ligands, enzymes with substrates, antibodies with antigens, and receptors with hormones, are characterized by a high degree of specificity. In these cases, specificity is achieved by the formation of noncovalent complexes in which at least one of the reacting species is macromolecular. Such biospecific interactions form the basis of a range of chromatographic techniques, including protein-binding assays. These assays for determining metabolites in biological samples are discussed at length.

A suitable method for the analysis of small samples in biological matrices is exemplified by the determination of ceftibuten in tracheal and bronchial secretion or in middle-ear fluid. Only a small volume ($<30 \mu L$) of these fluids is generally available. Trace–ultratrace LC–LC–UV assays are included for these samples. In addition, an LC–LC–MS method is discussed for determination of the drug in sputum samples, for which the other methods have proved unsuccessful.

A laboratory robotic system designed specifically for automating sample preparation and chromatographic analysis of drugs and their metabolites in biological fluids is discussed in detail. PyTechnology architecture, which combines new advances in both software and hardware, performs all necessary laboratory operations for the separation and analysis of compounds in the biological matrix.

The isolation and purification of large quantities of various types of glycerophospholipids are described. An explanation of why preparative HPLC is the technique of choice to isolate and purify large quantities of naturally occurring or synthetic phospholipids either by class or molecular species is provided. Additionally, an approach to CE method development for small-molecule pharmaceutical separations is demonstrated.

SATINDER AHUJA
Ciba–Geigy Corporation
Suffern, NY 10901

April 27, 1992

Chapter 1

Chromatography and Pharmaceutical Analysis

Satinder Ahuja

Development Department, Pharmaceuticals Division, Ciba–Geigy Corporation, Suffern, NY 10901

Pharmaceutical analysis can be simply defined as analysis of a pharmaceutical compound or drug. These analyses are needed in a variety of situations that range from an assay of the new chemical entity (NCE) in the presence of related compounds including optical isomers to complex determination of trace or ultratrace level of various related or transformation products. Chromatography plays a major role in pharmaceutical analysis as exemplified by the following:

> New Drug Development
> Purity/Impurity Analysis
> Separations of Isomers
> Support of Biotechnology Products
> Support of Toxicology Studies
> Biopharmaceutic/Pharmacokinetic Studies
> Metabolic Studies
> Clinical Studies
> Forensic Studies
> Diagnostic Studies
> Animal-derived Food Analyses
> Post-mortem Toxicology

Most of these topics are discussed in this book; discussion on the others may be found in a number of books (1-5).

New Drug Development

Development of new drug products mandates that meaningful and reliable analytical data be generated at various steps of new drug development (1). To assure the safety of a new pharmaceutical compound or drug requires that the new drug meet the established purity standards as a chemical entity and when admixed with animal feeds for toxicology studies or pharmaceutical excipients for human use. Ideally a high-quality drug product should be free of extraneous material and contain the labeled quantity of each ingredient. Furthermore, it should exhibit excellent stability throughout its shelf life. These requirements demand that the analytical methodology used be sensitive enough to carry out measurements of low levels of drugs and by-products. This has led to the development of

0097–6156/92/0512–0001$06.00/0

separation methods which are suitable for the determination of submicrogram quantities of various chemical entities.

Transformation Products

Transformation products or related products are impurities or by-products that originate during the synthesis of NCE, or they are degradation products arising during storage due to degradation of NCE or its interactions with the excipients or they are produced as metabolites in the body when the NCE is administered to a patient (1,4). There are a host of other places where these transformation products can be produced during the drug discovery process. Chromatography helps monitor these transformation products. The level to which they are controlled, however, relates to the potential hazard of a given transformation product. Isomeric impurities, discussed in the next section, may also be considered transformation products (5).

FDA Perspective on Isomeric Impurities

The current regulatory position of the Food and Drug Administration is described below with regard to the approval of racemates and pure stereoisomers (6). Circumstances in which stereochemically sensitive analytical methods are necessary to ensure the safety and efficacy of a drug are discussed. Regulatory guidelines for new drug application (NDA) are interpreted for the approval of a pure enantiomer in which the racemate is marketed, for the approval of either a racemate or a pure enantiomer in which neither is marketed, and for clinical investigations to compare the safety and efficacy of a racemate and its enantiomers. The basis for such regulation is drawn from historical situations (thalidomide, benoxaprofen) as well as currently marketed drugs (arylpropionic acids, disopyramide, indacrinone).

The primary regulatory focus of the Food and Drug Administration is on considerations of both clinical efficacy and consumer safety of a potential drug. Because the chiral environment found in vivo affects the biological activity of a drug, the approval of stereoisomeric drugs for marketing can present special challenges. The case of thalidomide is an example of a problem that may have been complicated by ignorance of stereochemical effects. The use of racemates can lead to erroneous models of pharmacokinetic behavior and to the potential for opportunities to manipulate pharmacologic activity. It is within the realm of technical feasibility to design experiments that will unambiguously answer the question whether or not a stereochemically pure drug is more effective and/or less toxic than the racemate.

In 1987, the FDA issued a set of guidelines on the submission of NDA where the question of stereochemistry was approached directly in the guidelines on the manufacture of drug substances (7). The FD&C Act requires a full description of the methods used in the manufacture of the drug, which includes testing to demonstrate its identity, strength, quality, and purity. Therefore, the submissions should show the applicant's knowledge of the molecular structure of the drug substance. For chiral compounds, this includes identification of all chiral centers. The enantiomer ratio, although 50-50 by definition for a racemate, should be defined for any other admixture of stereoisomers. The proof of structure should consider stereochemistry and provide appropriate descriptions of the molecular structure. The guidelines do not discuss conditions under which a determination of absolute configuration is desirable or essential. Obviously, it would be appropriate data for supporting the manufacture of optically pure drugs.

U.S. regulatory requirements demand that the bioavailability of the drug be demonstrated (8). When pharmacokinetic models differ between enantiomers, it seems obvious that establishing the bioavailability of the drug from a racemate is a much more complex task, which cannot be accomplished without separation of the enantiomers and investigation of their pharmacokinetics as individual molecular entities.

It is expected that the toxicity of impurities, degradation products, and residues from manufacturing processes will be investigated as the development of a drug is pursued. The same standards should, therefore, be applied to the enantiomeric molecules in a racemate (7).   Whenever a drug can be obtained in a variety of chemically equivalent forms (such as enantiomers), it makes sense to explore the potential in vivo differences between these forms.

Trace/Ultratrace Analyses

Frequently, trace/ultratrace analysis is required for monitoring transformation products and isomers (1, 4, 5). Trace analysis can be defined as analysis performed at parts per million (ppm) or microgram/gram (μg/g) level--an analytical landmark that was achieved approximately 30 years ago. Ultratrace (ultra; beyond what is ordinary) analysis can then be defined as analysis performed below ppm or μg/g level. Reliability of trace and ultratrace data requires that the data withstand interlaboratory comparisons. Frequently, meaningful intercomparisons are difficult because of the nature of uniformity of sample, the ease of contamination during sampling and analyses, and the limitations of analytical practices employed.

Separation Methods

Separation methods such as thin-layer chromatography (TLC), gas-liquid chromatography (GLC), and high pressure liquid chromatography (HPLC) are commonly used, and are discussed in this book. Of these methods, HPLC has revolutionized the field of separations (2, 3, 9). The compounds that once were considered too difficult to separate by TLC because of poor resolution or quantitation, or by GLC because of volatility, polarity, or thermal instability can be easily separated and quantified by HPLC within a short period of time. Furthermore, proteins and other macromolecules produced in biotechnology can also be resolved. HPLC offers selectivity to separate components with slight variations in structure or molecular weight. For compounds with the same molecular weight, the structural differences may involve no more than compounds that are mirror images, that is, optical isomers resulting from the presence of one or more asymmetric carbon atoms. Examples of high selectivity offered by HPLC or LC are discussed in this text. Also included are hyphenated methods such as GC/MS, LC/LC/UV, and LC/MS, LC/LC/MS, new methods such as capillary zone electrophoresis (CZE), and preparative separations of glycerophospholipids.

Applications

Numerous applications in the area of monitoring quality, biotechnology, diagnostic studies, biopharmaceutics/pharmacodynamics and clinical pharmacology are briefly discussed below to acquaint the reader with some of the exciting applications of chromatography.

Monitoring Quality. As mentioned before, impurities in pharmaceutical compounds originate mainly during the synthetic process from raw materials, solvents, intermediates, and by-products. Degradation products and contaminants of various types comprise some of the other sources of impurities. As a result, it is necessary to incorporate stringent tests to control impurities in pharmaceutical compounds. This fact is evident from the requirements of the Federal Food, Drug & Cosmetic Act and various pharmacopeias that provide tests for the control of specific impurities. A new drug development program should include an armamentarium of physicochemical tests to fully define the impurity of a pharmaceutical compound prior to performance of extensive pharmacologic and toxicologic studies. This is essential to assure that the observed toxicologic or pharmacologic effects are truly due to the compound of interest and not due to impurities.

The level to which any impurity should be controlled is primarily determined by its pharmacologic and toxicologic effects. This should include all impurities: those originating out of synthesis and those originating from other sources such as degradation. For example, penicillins and cephalosporins have been known to undergo facile cleavage of the β-lactam bond in aqueous solution. This is of special interest since some studies on penicillin have shown that their instability may effect possible reactions involved in penicillin allergy (10). The control of low level impurities is extremely important when a drug is taken in large quantities for therapeutic purposes or as a fad. Examples are the use of methotrexate (10-20 g) to treat neoplasia or the faddist use of vitamins, especially C.

Recombinant Products. The following recombinant-DNA-derived proteins have been approved for therapeutic use of clinical trials in USA:

> Insulin ("Humulin")
> Growth hormone ("Protropin")
> α-Interferon ("Referon")
> Tissue plasminogen activator
> γ-Interferon
> Tumor necrosis factor
> Interleukin-2

Particular attention must be paid to the detection of DNA in all finished biotechnology products because such DNA might be incorporated into the human genome and become a potential oncogene. The absence of DNA at the picogram-per-dose level must be demonstrated in order for the biotechnology products to be safe (11).

The isolation and purification of DNA and RNA restriction fragments are of great importance in the area of molecular biology today. These fragments are the product of site-specific digestion of larger pieces of DNA and RNA with enzymes called restriction endonucleases (12). The fragments may range in size from a few base pairs to tens of thousands of base pairs. The purification of restriction fragments is key to a number of processes, some of which are: cloning of proteins of peptides using engineered vectors (plasmids, phages, cosmids); sequencing of DNA; elucidation of the structure of the genome; and characterization of individual genes and gene effectors.

An ion-exchange column provides DNA and RNA separation within one hour, with resolution equivalent to that obtained using gel electrophoresis. Fragments are visualized using an on-line UV detector, and sample loadings from 500 ng to 50 μg have been applied successfully with recoveries approaching 100%. To date, DNA in the size range of 4 to 23,000 base pairs have been separated effectively. This technique could enable the molecular biologist to remove one of the most common bottlenecks encountered in a wide variety of experiments allowing fast, high-resolution, high-recovery purification of DNA and RNA.

Peptides, Proteins, and Antibodies. Reversed-phase liquid chromatography (RPLC) has recently gained acceptance as a powerful separation mode for peptides and proteins. In fact, RPLC is rapidly becoming the method of choice for resolving complex peptide mixtures from proteins cleavage reactions (for example, "peptide mapping" of CNBr and tryptic digests), discrimination of homologous proteins from different species, and separation of synthetic diasterioisomeric peptides (13). A typical application is peptide mapping of bacterial growth media. This application permits correlation of the presence of certain peaks with a medium's ability to support growth.

Molecular biologists are utilizing HPLC for the characterization and purification of protein, peptides, and antibodies (14). Analyses and separation of native and denatured protein structures with high resolution and minimal time have been accomplished on a variety of HPLC columns of single and mixed functionality.

Diagnostic Studies. Deficiency of xanthine-guanine phosphoribosyltransferase enzyme has been associated with Lesch-Nyhan syndrome as well as primary gout (15). The activity of the enzyme is determined by measurement of decrease of the substrate, hypoxanthine, and increase in the product, inosine-5'-monophosphoric acid. A major advantage of using HPLC for enzyme assays is that the simultaneous measurement of both substrate and product reduces the error due to interference from competing enzymes. Similarly the levels of hypoxanthine and uridine for colorectal cancer and inosine for gastric cancer have been found to be significantly higher in endoplastic mucosa than those in normal mucosa (P <0.05 with the paired t test) (16).

Biopharmaceutics/Pharmacodynamics/Clinical Pharmacology. Unless blood level studies of both the extent and rate of absorption are conducted in determining bioequivalence, a generic's inequivalence with respect to a metabolite may go undetected, and second-pass levels could be dangerously higher than indicated in the labeling (17).

Interesting investigations have been conducted to determine if the pharmacodynamics of the central nervous system stimulant pentylenetetrazol (PTZ) are altered in renal dysfunction (18). Female rats subjected to bilateral ureteral ligation (with sham-operated controls) or injected with uranyl nitrate (with saline-injected controls) were infused intravenously with PTZ until the onset of either a minimal (myoclonic jerk) or maximal (tonic hindlimb extension) seizure. Neither chemically nor surgically induced renal dysfunction caused a change in the concentrations of PTZ in cerebrospinal fluid serum or the brain at onset of minimal seizures. When PTZ was infused to the onset of maximal seizures, the rats with chemically induced renal dysfunction required higher concentrations, whereas the ureter-ligated rats convulsed at lower concentrations of PTZ than did the corresponding control animals. Thus, the effects of experimental renal dysfunction on the convulsant action of PTZ are dependent on both the disease model and the end point used for the pharmacodynamic measurement. Apparently, renal dysfunction did not affect the PTZ-induced seizure threshold, but inhibited the spread of seizures. The increased sensitivity of ureter-ligated rats may be due to their pronounced retention of water since water loading is known to increase seizure susceptibility. The effect of experimental renal dysfunction on the convulsant activity of PTZ was examined by a gas chromatographic method for studying pharmacodynamics of PTZ seizure in rats. Nitrogen-phosphorus detector was used to provide a detection limit of about 0.5 μg/mL.

Catecholamines are biologically active derivatives of the amino acid tyrosine that are produced by cells of the central nervous system (CNS) and the adrenal medulla. In the central nervous system, norepinephrine and dopamine function as neurotransmitters. Even subpopulations of cells within a given area have been shown to contain differences in levels and types of neurotransmitters (19). Highly sensitive methods will help us better understand brain function.

To summarize, this book deals with analysis of both low and high-molecular weight compounds and addresses some of the new developments in the area of chromatography and pharmaceutical analysis.

References

1. S. Ahuja, Ultratrace Analysis of Pharmaceuticals and Other Compounds of Interest, Wiley, New York, 1986.
2. S. Ahuja, Ultrahigh Resolution Chromatography ACS Symposium Series 250, American Chemical Society, Washington D.C., 1984.
3. S. Ahuja, Selectivity and Detectability Optimization in HPLC, Wiley, New York, 1989.
4. S. Ahuja, Trace and Ultratrace Analysis by HPLC, Wiley, New York, 1991.
5. S. Ahuja, Chiral Separations by HPLC, ACS Symposium Series 471, American Chemical Society, Washington D.C., 1991.
6. W. H., DeCamp, Chirality, 1, 2 (1989).
7. Guidelines for Submitting Supporting Documentation in Drug Applications for the Manufacture of Drug Substances. Office of Drug Evaluation and Research (HFD-100), Food and Drug Administration, Rockville, MD, 1987.
8. Code of Federal Regulations (FGR), Title 21, Government Printing Office, Sec. 314.5 (d) (3), Washington, D.C., 1988.
9. S. Ahuja, Chromatography and Separation Chemistry ACS Symposium Series 297, American Chemical Society, Wasington, D.C., 1986.
10. T. Yamana and A. Tsuji, J. Pharm. Sci., 65, 1563 (1976).
11. F. M. Bogdansky, Pharm. Technol. 72 (Sept. 1987).
12. M. Merim, W. Warren, C. Stacey and M. E. Dwyer, Waters Bulletin.
13. J. M. DiBussolo, Am. Biotechnology, p. 20, June 1984.
14. R. H. Guenther, J. Cocuzza, H. D. Gopal, and P. F. Agris, Am. Biotechnology, p. 22, Sept./Oct. 1987.
15. A. P. Halfpenny and P. R. Brown, J. Chromatogr., 199, 275 (1980).
16. K. Nakamo, K. Shindo, T. Yasaka, and H. Yamamoto, ibid, 332, 127 (1985).
17. J. Dickinson, Pharm. Technol., p. 14, March 1989.
18. I. Ramzan and G. Levy, J. Pharm. Sci., 78, 142 (1989).
19. J. Warnska, Waters Chromatography Bulletin.

RECEIVED July 13, 1992

Chapter 2

Chromatography and New Drug Development

Satinder Ahuja

Development Department, Pharmaceuticals Division, Ciba–Geigy Corporation, Suffern, NY 10901

Development of new drugs requires that meaningful and reliable analytical data be generated at various steps of development. To assure the safety of a new pharmaceutical compound or drug requires that the drug meet the established purity standards as a chemical entity and when admixed with pharmaceutical excipients and animal feeds. This demands that the analytical methodology used be sensitive enough to carry out measurements of low levels of the drug or its transformation products. These requirements necessitate that the developed methodologies should be selective, sensitive, and provide high resolution. The discovery of new methodologies that are suitable for determination of trace to ultratrace levels of various compounds are highlighted, and matrix effects on detection and quantification at these levels are discussed.

Discovery of new drugs mandates that meaningful and reliable analytical data be generated at various steps of development. This need has led to the development of new chromatographic methodologies that are suitable for the determination of trace/ultratrace quantities of various chemical entities. The development and application of these methodologies to assure purity, stability, and quality in the new drug development process are discussed below.

DISCOVERY OF NEW COMPOUNDS

Discovery of new drugs generally occurs via synthesis; however, chemists can discover potential drugs by several alternate routes. For example, chromatography provides an excellent means of discovery of new compounds since compounds present even at ultratrace levels can be resolved from related compounds (1). One approach entails separation of potential compounds resulting as by-products in the synthesis of the target compound, which cannot be resolved by normal inefficient crystallization techniques. Many of the by-products frequently have physicochemical properties and a carbon skeleton similar to the target compound with substituent(s) differing in position or functionality. Alternatively, they may be isomeric compounds which differ only as a

0097–6156/92/0512–0007$06.00/0

mirror image. Since it is not possible to theorize all by-products, some unusual compounds can be isolated and characterized with this approach.

A more selective approach is based on changes brought about in a chemical entity with reactions such as hydrolysis, oxidation, or photolysis. These reactions are frequently run to evaluate stability. In this case, several theorized, new, and old compounds are produced. An innovative chromatographer can resolve and characterize the theorized as well as untheorized new compounds.

Another interesting approach depends on characterization of various degradation products produced in the matrices used for pharmaceutical products. The compounds thus produced or those produced during metabolic studies in humans can be resolved by chromatography, and their structure determined by techniques such as elemental analysis, IR, NMR, or mass spectrometry. Two well-known examples of metabolites that became drug products are Tandearil® and Pertofrane®. These compounds resulted from oxidation and demethylation respectively of the parent compound.

Discovery of a number of compounds during our investigations is described in this paper, some of which were investigated as potential drugs. Discussed in detail below is the interrelationship of chromatography and new drug development.

ANALYSIS

Innovative analytical chemistry is requisite for studies on purity, stability, and quality of drugs. Frequently these analyses have to be performed at low levels. Analyses performed at parts per million (ppm) or microgram level are generally defined as trace analyses--an analytical landmark that was attained approximately thirty years ago. Analyses performed below ppm level, or submicrogram analyses, may be defined as ultratrace analyses (2). The methods useful for ultratrace analysis by chromatography are shown in Table I.

Table I. Chromatographic Methods Frequently Used for Ultratrace Analyses

Method	Minimum Amount Detected, g
Liquid chromatography	
Ultraviolet detection	10^{-11}
Fluorescence detection	10^{-12}
Electrochemical detection	10^{-12}
Gas chromatography	
Flame ionization	10^{-12} - 10^{-14}
Electron capture	10^{-13}
Combination techniques	
Liquid chromatography/Mass spectrometry	10^{-12}
Gas chromatography/Mass spectrometry	10^{-12}
Electron capture--negative ionization mass spectrometry	10^{-15}

In scientific literature, detectabilities are frequently referred to in a variety of units. It is recommended that a common unit such as grams be used for reporting analytical data (Table II) to permit better comparison of the detectabilities of the methods. A comparative view of the analytical quantities in terms of gram (g), percent, and parts per million is provided in Table II.

Table II. Preferred and Commonly Used Analytical Units

Preferred* Analytical Units		Commonly Used Units When Present/g	
g	Common Name	%	One Part Per--
1×10^{-6}	microgram (mg)	0.0001	million
1×10^{-9}	nanogram (ng)	0.0000001	billion
1×10^{-12}	picogram (pg)	0.0000000001	trillion
1×10^{-15}	femtogram (fg)	0.0000000000001	quadrillion
1×10^{-18}	attogram (ag)	0.0000000000000001	quintillion

* by author

Furthermore, to allow better application of methodologies across various disciplines, it is recommended that the following important analytical parameters be reported for each method:

> Amount present in original sample per mL or g (APIOS)
> Minimum amount detected in g (MAD)
> Minimum amount quantitated in g (MAQ)

In addition, data such as accuracy, precision, linearity, and specificity of the methodology should be provided.

Sampling and Sample Preparation. The sample used for analysis should be representative of the "bulk material." The major considerations are (3):

1. Determination of the population or the "whole" from which the sample is drawn.
2. Procurement of a valid gross sample.
3. Reduction of the gross sample to a suitable sample for analysis.

It is desirable to reduce the analytical uncertainty to a third or less of sampling uncertainty (4). Poor analytical results can also be obtained because of reagent contamination, operator errors in procedure or data handling, biased methods, etc. These errors can be controlled by proper use of blanks, standards, and reference samples. It is also important to determine the extraction efficiency of the method.

Frequently, preconcentration of the analyte is necessary because the detector used for quantitation might not have the necessary detectability, selectivity, or freedom from matrix interferences (5). During this step, significant losses can occur due to very small volume losses on glass walls of recovery flask or disposable glass pipets and other glassware.

Method Validation and Interlaboratory Variations. Statistically designed studies have to be performed to determine accuracy, precision, and selectivity of the methodology used for analyses. Reliability of low-level data requires that the data withstand interlaboratory comparisons. Frequently, meaningful comparisons are difficult because of the nature and uniformity of sample, the ease of contamination during sampling and analysis, and the variety and limitations of analytical practices employed.

Optimum Methodology Selection. It is desirable to select the optimum methodology, as exemplified below. Analytical methods were needed to evaluate a starting material for the synthesis of one of our drugs (6). This material has the following structure:

$$C_6H_5CH_2O \longrightarrow \longrightarrow COCH_3$$
$$C_6H_5CH_2O$$

A batch (ARD 36792) showed purity of 96.3% by gas chromatography (GC) and 93.3% by titrimetry after oximation with hydroxylamine hydrochloride. Thin layer chromatography (TLC) performed with an optimum system (90 CCL_4/5 acetone/2 HOAc) showed two impurities totaling $\approx 8\%$. This suggested titrimetry in combination with TLC would be adequate to control purity of this material. To check this hypothesis, a high performance liquid chromatographic (HPLC) method was developed (4 methanol/1 H_2O was used as the mobile phase with a C_{18} column, and detector set at 225 nm).

No impurities were detected in this sample by TLC; however, HPLC showed three impurities totaling 0.06%. The absolute purity of this sample by differential scanning calorimetry (DSC) was 99.7 mole %. Further comparisons on a sample from an alternate source revealed that HPLC was the method of choice since it could help differentiate samples that appeared comparable by DSC: the sample from an alternate source was found to be more pure (only two impurities $\approx 0.02\%$) than the reference sample (five impurities $\approx 0.12\%$).

Purity. Impurities in pharmaceutical compounds originate mainly during the synthetic process from raw materials, solvents, intermediates, and by-products. Degradation products and contaminants of various types comprise some of the other sources of impurities. As a result, it is necessary to incorporate stringent tests to control impurities in pharmaceutical compounds. This fact is evident from the requirements of the Federal Food, Drug & Cosmetic Act and various pharmacopeias that provide tests for the control of specific impurities.

A new-drug development program should include an armamentarium of physicochemical tests to fully define the purity of a pharmaceutical compound prior to performance of extensive pharmacologic and toxicologic studies (7). This is essential to assure that the observed toxicologic or pharmacologic effects are due to the compound of interest and not due to impurities.

Analytical methods that can control impurities to ultratrace levels are available (2, 8-10); however, the level to which any impurity should be controlled is determined primarily by its pharmacologic and toxicologic effects. This should include all impurities: those originating out of synthesis and those originating from other sources such as degradation (11). For example, penicillins and cephalosporins have been known to undergo facile cleavage of the β-lactam bond in aqueous solution. This is of special interest since some studies on penicillins have shown that their instability may effect possible reactions involved in penicillin allergy (12). The control of low level of impurities is extremely important when a drug is taken in large quantities for therapeutic purposes or as a fad. Examples are the use of methotrexate (10-20 g) to treat neoplasia or the faddist use of vitamins, especially vitamin C.

In general, impurities are controlled down to 0.1% level for pharmaceutical compounds unless carcinogenicity or toxicity studies dictate lower levels. Hence from a practical standpoint, determinations from ≥0.1% down to ultratrace levels are of interest to a pharmaceutical analyst. To fully evaluate impurities, it is important that they be resolved from the main component by separation methods such as chromatography. Schemes that group compounds on the basis of acidity or basicity may be used for preliminary separations. The main impurities of interest then should be isolated and/or synthesized and their limits established on the basis of various biological tests. The fact that no specifications for impurities exist or that they can vary between pharmacopeias suggests that this is not being done uniformly today. For example:

- No impurities limits are given in the United States Pharmacopeia (USP) for indomethacin, even though it can be synthesized by several routes.

- USP titrimetric assay for hydrochlorothiazide cannot distinguish it from degradation and in-process impurities such as chlorothiazide and 4-amino-6-chloro-1,3-benzenedisulfonamide.

- USP reference-grade material of ethotrexate gives a purity figure of only 86%, based on a chromatographic method (13).

- USP allows dihydroquinidine concentration of up to 20% by a limit test in quinidine.

- USP has no specifications for impurities in meprobamate, and the British Pharmacopeia (BP) controls related substances to 1%.

- BP allows fivefold concentration (0.05%) of 2-amino-5-chlorobenzophenone impurity in chlordiazepoxide as compared to the USP.

A variety of methods can be used for monitoring purity (see Table I); however, it is not always necessary to use the most elaborate or most expensive method. TLC, for example, provides a simple means of controlling impurities in imipramine and desipramine. Two solvent systems with slightly different compositions (Table III) are used for monitoring impurities by TLC on Silica gel G plate with 254 nm fluorescent indicator (14). Due to safety considerations, it may be desirable to replace benzene with another solvent such as toluene. The Rf and minimum detectability values of the impurities are given in Table III.

Table III. TLC of Imipramine and Desipramine

Compound	Rf	Minimum Detectability μg
Iminodibenzyl	0.81	0.06
Imipramine	0.67	0.03
Unknown	0.26	0.03
Desipramine	0.36	0.03

Imipramine - Benzene:ethyl acetate:absolute ethanol:conc. ammonium hydroxide (50:50:15:3)

Desipramine - Benzene:ethyl acetate:absolute ethanol:conc. ammonium hydroxide (50:50:20:1)

SOURCE: Adapted from ref. 14.

The unknown impurity has been characterized by comparison with synthesized material and has the following structure:

Iminodibenzyl is a synthetic precursor and may also result from breakdown of the drug. Desipramine (Pertofrane) is a metabolite of imipramine (Tofranil®), which has been used for the same indications as the parent drug. Further discussion on purity may be found under the following sections on Stability and Quality.

STABILITY

There are legal, moral, economic, and competitive reasons, as well as those of safety and efficacy, to monitor, predict, and evaluate drug product stability (11). However, stability can and does mean different things to different people or to the same people at different times, even those in pharmaceutical science and industry. The fact remains that stability of the product is of great concern to those involved with analytical, production, marketing, and distribution departments, and to the physician, pharmacist, and patient. This concern is manifested by the use of storage legends, expiration dates, protective packaging, and dispensing directions. Furthermore, from a regulatory viewpoint, one should assure that the product is of the "quality, strength, purity, and identity" that it is purported to be throughout the time it is held or offered for sale.

One cannot monitor stability, determine the reaction rate, or investigate any mechanism without an analytical measurement. Hence, the pharmaceutical analyst is primarily involved in stability because he or she must develop a method that will quantitatively determine the drug in the presence of, or separate from, the transformation product(s). This determination is required to assure that the drug has not undergone change. To select the appropriate method(s), the analyst should have a thorough knowledge of the physicochemical properties of the drug, including an understanding of the routes by which a drug can be degraded or transformed.

Kinetics. Most degradation reactions of pharmaceuticals occur at finite rates and are chemical in nature. These reactions are affected by conditions such as solvent, concentration of reactants, temperature, pH of the medium, radiation energy, and presence of catalysts. The manner in which the reaction rate depends on the concentration of reactants describes the order of the reaction. The degradation of most pharmaceuticals can be classified as zero-order, first-order, or pseudo-first-order, even though these compounds may degrade by complicated mechanisms and the true expression may be of higher order or be complex and noninteger.

An understanding of the limitations of the experimentally obtained heat of activation values is critical in stability prediction. For example, the apparent heat of activation of a pH value where two or more mechanisms of degradation are involved is not necessarily constant with temperature. Also, the ion product of water, pk_w, is temperature-dependent, and $-\Delta H_a$ is approximately 12 kcal, a frequently overlooked factor that must be considered when calculating the hydroxide ion concentration. Therefore, it is necessary to obtain the heats of activation for all bimolecular rate constants involved in a rate-pH profile to predict degradation rates at all pH values for various temperatures.

The basic kinetic effects are important to an understanding of the reaction and of possible adverse, practical effects. For example, addition of an inert salt such as sodium chloride to adjust isotonicity can affect the reaction rate as a primary salt effect. Buffers used to control pH are also ionic species and can exert a primary salt effect. In addition, they exert a secondary salt effect and act as catalysts. Sulfite salts are frequently added as antioxidants, but they can form addition products with the active ingredient, or they can act as catalysts.

Organic solvents such as alcohol are generally used for solubilization; the concentration of the organic solvent can affect the dielectric constant of the solvent and thus influence the degradation rate of the active ingredients. The preservative used to inhibit bacterial growth, or other pharmaceutical aids may decompose, and decomposition products may, in turn, influence the decomposition rate of the active ingredients by one or more of the means discussed previously.

Complex reactions, including reversible reactions, consecutive reactions, and parallel reactions are occasionally encountered in the decomposition of pharmaceuticals. Numerous examples of hydrolytic reaction, oxidation, catalysis, miscellaneous reaction and effect of pharmaceutical aids on stability of A.I. were summarized in a review article (11). An example of metal catalysis at ultratrace levels is discussed below.

Metal Catalysis at Ultratrace Levels. 1,3-Benzenediols substituted with an amino-alkyl moiety are of significant pharmacologic interest. Stability investigations on substituted 5-amino-ethyl-1,3-benzendiol sulfate (AEB), under various exaggerated conditions, revealed that AEB is susceptible to degradation in aqueous solution in the presence of metals (15, 16). Metal cations such as copper, iron, and calcium have been shown to accelerate degradation of AEB under an oxygen atmosphere, with concomitant discoloration. The effectiveness of these metals in terms of AEB degradation is in the following order: $CU^{+2} > Fe^{+3} > Ca^{+2}$. Copper ($Cu^{+2}$) effectively catalyzes AEB degradation down to the ten parts per billion (ppb) level in the presence of oxygen.

A significant increase in the rate of AEB degradation occurs when the concentration of Cu^{+2} is increased from 10 ppb to 1,000 ppb (Table IV). The increase in the rate of degradation is less pronounced at higher concentrations such as 1 to 10 ppm, on parts per million basis.

Table IV. Determination of Optimum Concentration of Copper for Kinetic Studies in Oxygen Atmosphere

Copper Concentration	% AEB Remaining	
(ppm)	After 18 Hours at 90°C	After 41 Hours at 90°C
10	35.2	8.7
5	45.0	17.7
1	64.78	31.2
0.5	69.9	39.6
0.01	89.8	59.1

SOURCE: Adapted from ref. 15.

Kinetic studies have been performed at 1 ppm Cu^{+2} concentration to determine the effect of pH (in the range of interest) and temperature on degradation of AEB by monitoring the degradation rate with a selective high performance liquid chromatographic method (Table V). The method entails separation on μBondapak C_{18} column (30 cm x 4.6 mm i.d.) with a mobile phase containing 0.0028M 1-octanesulfonic acid in 36 methanol:64 water:1 acetic acid followed by detection at 280 nm.

Table V. Kinetic Studies of AEB Solution (1 mg/mL) in an Oxygen Atmosphere (1 ppm Cu)

Time	% AEB Remaining								
(hrs)	pH 3.0			pH 4.0			pH 5.0		
	90°C	70°C	50°C	90°C	70°C	50°C	90°C	70°C	50°C
7	88.5	--	--	95.9	--	--	96.9	--	--
24	42.5	85.3	--	93.8	98.6	--	96.5	98.4	--
48	21.7	56.8	101	86.4	98.7	99.3	91.2	101	100
120	2.5	--	--	31.0	--	--	64.0	--	--
123	--	28.2	--	--	--	--	--	--	--
144	--	--	63.8	--	63.8	96.0	--	--	99.1
164	--	19.3	--	--	--	--	--	--	--
167	--	--	--	23.4	--	--	--	--	--

Continued on next page

Time	% AEB Remaining								
	pH 3.0			pH 4.0			pH 5.0		
(hrs)	90°C	70°C	50°C	90°C	70°C	50°C	90°C	70°C	50°C
213	--	12.4	51.1	--	49.7	--	--	--	--
216	--	--	--	--	--	--	40.0	--	--
332	--	--	31.9	--	--	74.2	--	97.1	--
452	--	--	23.1	--	24.0	61.0	--	--	--
787	--	--	--	--	--	47.7	--	66.5	--
1147	--	--	--	--	--	--	--	48.3	--
1892	--	--	--	--	--	--	--	--	94.0

Table V. Continued
(1 ppm Cu) (15)

SOURCE: Adapted from ref. 15.

The results of these studies (Table V) show that the pH of the solution has greatest effect on AEB degradation. The rate of degradation is fastest at pH 3 and slowest at pH 5. At pH 4 and pH 5 the degradation reaction appears to proceed in two steps (k_1, k_2) instead of an apparent single step at pH 3. The postulated second step (k_2) is relatively fast. At pH 3 (90°C), k_1 is indistinguishable from k_2, and its calculated value is 17 times higher than that at pH 5 (Table VI). It should be noted that since k_1 values are small and based on few data points, these extrapolations must be used judiciously. Based on k_2 data at 90°C, it would appear that a solution at pH 5 is ≈ 6 times more stable than one at pH 3.

**Table VI. Calculated Apparent First-Order Rate Constants for
Kinetic Study
(Data in Table V)**

Temperature	pH	k_1 (hour^{-1})	k_2 (hour^{-1})
90°C	3.0	*	0.0318
	4.0	0.0029	0.0140
	5.0	0.0019	0.0049
70°C	3.0	*	0.0103
	4.0	0.0003	0.0042
	5.0	0.00008	0.0008
50°C	3.0	--	0.0041
	4.0	0.00026	0.0014
	5.0	0.00003	--

*$k_1 = k_2$

SOURCE: Adapted from ref. 15.

Two degradation products (II and III; Figure 1) were prepared in our laboratory (17). Their presence could not be demonstrated in the solutions degraded under exaggerated conditions; however, very small peaks for resorcylic acid (VII) and 3,5-dihydroxy benzaldehyde (VI) were found by HPLC. This suggests that oxidative degradation is favored (Figure 1).

Effect of Instability of One Compound on Another. A non-marketed phenylbutazone product containing prednisone and alkalizer showed a large increase in prednisone assay values with blue tetrazolium (BT) reagent under accelerated conditions. An investigation by TLC revealed the presence of four impurities (II, III, IV, and VI in Figure 2) none of which produced color with BT reagent. A BT spray reagent and a new TLC system were developed in our laboratories (18) to track the degradation product(s) responsible for the problem. Two new transformation products were characterized (V & VII). Compound VII was found to be responsible for interference in the prednisone assay.

Importance of Selective Methodology. Two capsule formulations containing homatropine methylbromide (HMB), antacids, and other ingredients were investigated for HMB stability under accelerated conditions. TLC followed by spraying with Dragendorff reagent revealed the presence of a purple spot. The purple spot was attributed to degradation of HMB to methyltropinium bromide or tropine methyl bromide (TMB). This suggested that the Dragendorff reaction could be useful for monitoring stability of HMB. The method would be stability indicating because TMB does not yield any precipitate with Dragendorff reagent, and therefore would not interfere in the assay. However, it was found necessary to stabilize Dragendorff calorimetric reaction (19) by the addition of methanol to an acetone solution of HMB-iodobismuthate complex in the ratio of 1:1. The stabilization was accompanied by hypsochromic shift from 474 to 382 nm. Stability data on several samples showed that a loss as high as 58% degradation of HMB could go undetected by the titrimetric method (20). Further discussion on selective methodology is given in the next section.

SUPPORT OF TOXICOLOGY STUDIES

It was necessary to conduct a long-term toxicity study to evaluate the safety of a potential drug with the following structure (6, 20).

Molecular Weight: > 400
M.P.: 224 °C (decomp.)
Solubility: Very slightly soluble in water. Almost insoluble in acidic and
                  basic solution. Soluble in methanol.
pKa: 5.0

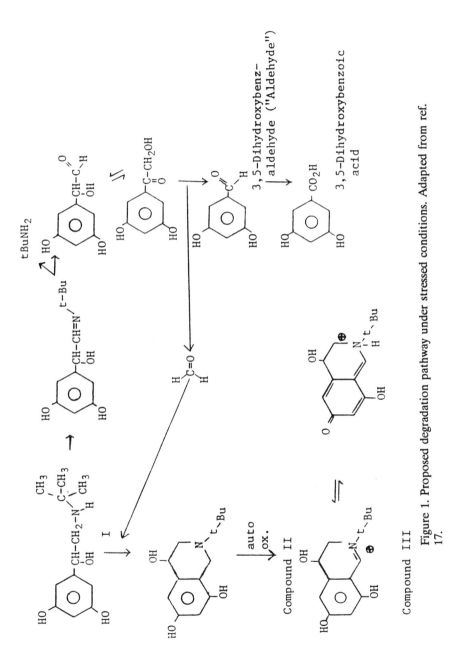

Figure 1. Proposed degradation pathway under stressed conditions. Adapted from ref. 17.

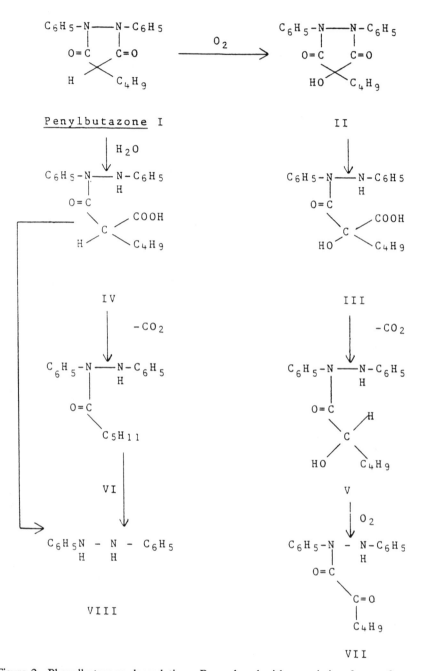

Figure 2. Phenylbutazone degradation. Reproduced with permission from reference 1. Copyright Wiley 1985.

This compound was admixed with rat feed (Purina Laboratory Chow containing a minimum of 23% protein, 4.5% fat, and maximum 6% fiber) with the following composition: Meat and bone meal, dried skimmed milk, wheat germ meal, fish meal, animal liver meal, dried beet pulp, ground extruded corn, ground oat groats, soybean meal, dehydrated alfalfa meal, cane molasses, animal fat preserved with BHA (butylated hydroxyanisole), Vitamin B_{12} supplement, brewers' dried yeast, thiamin, niacin, vitamin A supplement, D activated plant sterol, vitamin E supplement, dicalcium phosphate, iodized salt, ferric ammonium citrate, zinc oxide, manganous oxide, cupric oxide, ferric oxide, and cobalt carbonate.

A gas-liquid chromatographic method was first investigated for analysis of this compound. Electron capture detector provided detectability down to five pg for the active component; however, investigations revealed degradation was occurring at the injection port. This clearly shows that the developed method must not only be sensitive but should be free from interference from the degradation products and/or impurities, i.e., it must be selective (21). A selective HPLC method was then developed to circumvent this problem. The essential details of the method follow:

> Column: μBondapak C_{18}
> Precolumn: 6 cm x 2.5 mm i.d. C_{18} Corasil
> Solvent System: 70 MeOH/33 H_2O/1 HOAc
> Flow Rate: 0.8 mL/min
> Methodology:
> Sample + 5 mL 1 $\underline{N}$ HCl + 20 mL EtOAc.
> Shake. Centrifuge. Inject 25 μL of EtOAc layer.

Selectivity of the HPLC method is given in Table VII. The discovery of the "Hydroxy Compound" (oxidation product) is primarily attributed to selectivity and detectability offered by chromatography. It is also possible to separate the ketosulfone and RNH_2 (hydrolysis products) from the parent compound.

TABLE VII. Selectivity of HPLC Method

	Retention Time (minutes)
"Ketosulfone"	5.0
RNH_2	6.9
RNCO	10.1
"Hydroxy compound"	13.0
Parent compound	18.0

Analysis of 20-week-old samples revealed that all samples degraded essentially linearly. These lines, when extrapolated to zero time, indicated that essentially 100% of labeled amount was present at the start of the study. These investigations on the cause of degradation helped to minimize degradation and yield suitable feed mixtures for the toxicity study.

DETERMINATIONS OF UNSTABLE COMPOUNDS

The problems discussed earlier for ultratrace analyses are further compounded when one is dealing with compounds such as hydrazobenzene and azobenzene. Hydrazobenzene is known to be an unstable compound; it oxidizes easily to azobenzene and other compounds, and has $t_{1/2}$ of 15 minutes in waste-water (22). Azobenzene, on the other hand, can isomerize or sublime even at 30°C (23). Discussed below are methods developed in our laboratory which circumvent some of the problems encountered with them (24).

A review of the literature revealed that a normal-phase HPLC method has been used for the analysis of hydrazobenzene and azobenzene (25). The method entails extraction of these compounds into n-hexane from 1$\underline{N}$ NaOH followed by analysis on Partisil-10 PAC column with a mobile phase containing 2.5% absolute ethanol. This method suffers from the following shortcomings:

a.   Hydrazobenzene and azobenzene show significant instability in 1$\underline{N}$ NaOH (Table VII).

b.   Azobenzene can isomerize into cis- and trans- isomers. Their separation is not demonstrated or accounted for in the method.

c.   The parent compound can degrade directly or indirectly into hydrazobenzene and azobenzene (Figure 2).

d.   Selectivity of transformation products given in Figure 2 is not demonstrated.

To assure that the methodology would be reliable at $\simeq$ 10 ppm, suitable methods were developed for detecting these compounds at levels $\leq$ 1 ppm, i.e., ultratrace levels.

The method entails taking a sample weight anticipated to contain ~ 10 ppm of hydrazobenzene or azobenzene and shaking it with 30 mL of pH 9.2 THAM buffer. This is followed by extraction with 10 mL of n-hexane. After centrifugation, 5 mL of the n-hexane layer is evaporated to dryness at room temperature, with nitrogen; the residue is solubilized in 1.0 mL of acetonitrile. Twenty-five microliters are immediately injected into HPLC equipped with Partisil 10 μ C$_8$ column (25 cm x 4.6 mm) and a dual channel detector (254 and 313 nm). Elution is carried out with a mobile phase composed of acetonitrile:acetate buffer, pH 4.1 (11:14). Both hydrazobenzene and azobenzene standards are treated similarly.

Investigations revealed that the optimum pH for extraction for both hydrazobenzene and azobenzene is 9.2. At this pH, these compounds can be easily extracted from the parent compound and are also quite stable (Table VIII). The cis- and trans- isomers of azobenzene and hydrazobenzene can be resolved well with the reversed-phase HPLC method (Figure 3). Previous investigations had confirmed the selectivity of this method as it resolves compound I (for structure, see Figure 2) from other transformation products (26). Data on spiked samples are given in Table IX. Average recoveries of 91% and 114% were obtained for hydrazobenzene and azobenzene respectively with relative standard deviation (R.S.D.) of 3-11%. The methods were found useful for quantitating $\leq$ 1 μg/g of these compounds with respect to the parent compound (MAD = 6-7 ng). The high recovery obtained for azobenzene is largely attributable to conversion of hydrazobenzene to azobenzene (~9%).

Table VIII. Stability of Hydrazobenzene and Azobenzene

| Medium | Time | % Loss | |
		Hydrazobenzene	Azobenzene
0.1N NaOH	30 minutes	82.9%[a]	4.6%[b]
pH 9.2 buffer	30 minutes	0.9%[c]	None found[d]

Original concentration in 10% acetonitrile:
a 11.8 μg hydrazobenzene/mL; b 15.7 μg azobenzene/mL; c 3.55 μg hydrazobenzene/mL; d 2.59 μg azobenzene/mL
SOURCE: Adapted from ref. 24.

Table IX. Recovery Data of Spiked Samples

APOIS: ≤ 10 μg/g of Parent Compound

Sample	Hydrazobenzene Found	Azobenzene Found
Parent compound	89.0 ± 8.6% (n = 7)	123 ± 2.6% (n = 6)
Capsules	89.6% ± 10.8% (n = 5)	98.7 ± 10.2% (n = 3)
Tablets	95.8 ± 5.4% (n = 3)	121 ± 4.2% (n = 3)
Average	91%	114%
R.S.D.	±5 = 11%	±3 - 10%
MAD (μg)	0.006 (6 ng)	0.007 (7 ng)
MAQ (μg/g)	≤ 1	≤ 1

SOURCE: Adapted from ref. 24.

Separation of Chiral Compounds

Molecules that relate to each other as an object and its mirror image that is not superimposable are enantiomers, or chiral (from the Greek word, cheiro, meaning hand); they are like a pair of hands. Stereoisomers are isomeric molecules with identical constitution but a different spatial arrangement of atoms. The symmetry factor classifies stereoisomers as either enantiomers, as defined above, or diastereoisomers. A pair of enantiomers is possible for all molecules containing a single chiral carbon atom (one with four different groups attached). Diastereoisomers are basically stereoisomers that are not enantiomers of each other. Although a molecule may have only one enantiomer, it may have several diastereoisomers. However, two stereoisomers cannot be both enantiomers and diastereoisomers of each other simultaneously.

The importance of determining the stereoisomeric composition of chemical compounds, especially those of pharmaceutical importance, cannot be overemphasized (27, 28). Dextromethorphan provides a dramatic example in that it is an over-the-counter antitussive, whereas levomethorphan, its stereoisomer, is a controlled narcotic.

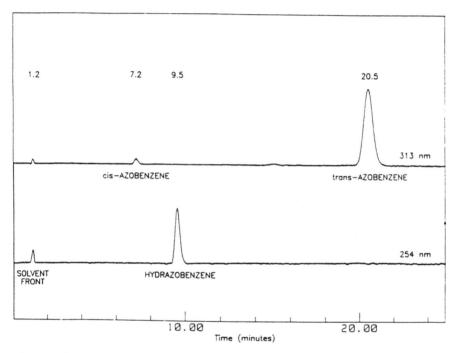

Figure 3. Chromatograms of *cis*- and *trans*-azobenzene and hydrazobenzene. Injection of 0.05 μg sample of each compound monitored at 313 and 254 nm. Adapted from ref. 24.

Likewise, it has been reported that the teratogenic activity of thalidomide may reside exclusively in the (S)-enantiomer (29). Less dramatic examples abound; 12 of the 20 most prescribed drugs in the USA and 114 of the top 200 possess one or more asymmetric centers in the drug molecule (30). About half of the 2050 drugs listed in the U.S. Pharmacopeial Dictionary of Drug Names contain at lease one asymmetric center, and 400 of them have been used in racemic or diastereomeric forms (31). The differences in the physiologic properties between enantiomers of these racemic drugs have not yet been examined in many cases, probably because of difficulties of obtaining both enantiomers in optically pure forms. Some enantiomers may exhibit potentially different pharmacologic or toxicological activities, and the patient may be taking a useless, or even undesirable, enantiomer when ingesting a racemic mixture. To ensure the safety and effect of currently used and newly developing drugs, it is important to resolve both enantiomers. One such example is discussed below.

Baclofen (4-amino-3(p-chlorophenyl)butyric acid) is a synthetic antispastic agent or muscle relaxant. The synthesis yields the racemic form which is used therapeutically (32). Based on the results of an animal study, the ℓ-form is much more active than the d-form (Table X). The depression of different reflexes in cats and protection against electroshock-induced convulsions in mice are demonstrable for ℓ-baclofen only. However, it should also be noted that the baclofen toxicity exists mainly in the ℓ-enantiomer. This suggests that it would be desirable to resolve d- and ℓ- isomers. The results of our investigations to resolve the d- and ℓ-enantiomers with HPLC are discussed below.

Table X. Activity of Baclofen Enantiomers

Property	(-)-ℓ-baclofen	(+)-d-baclofen
Specific rotation (1% solution)	-1.4°	1.4°
Reflex inhibition (i.v. cat)	100% at 1-3 mg/kg	0% at 30 mg/kg
Protection E-shock (p.o. mouse)	50% at 60 mg/kg	0% at 100 mg/kg
Acute toxicity (p.o. mouse; 1000 mg/kg)	9/10	0/10

SOURCE: Adapted from ref. 33.

Previous investigators have utilized derivatization and/or chiral mobile phases:

- Weatherby et al (34) used a chiral mobile phase to separate tritium-labeled enantiomers.

- Smith and Pirkle reported separation of DNB derivatives of the methyl ester of Baclofen (35).

- Wuis et al derivatized Baclofen with a chiral adduct of o-phthaladelhyde (36).

●   Spahn et al carried out chiral derivatization on the butyl ester of
       Baclofen with S(+) naproxen chloride (37).

A simple method based on ligand exchange on Chiralpak WH column (Daicel)
was developed. An aqueous solution of cupric sulfate (0.25 mM) is used as a mobile
phase at a flow rate of 1.5 mL/minute. The method provides resolution of d-and ℓ-
forms. This was confirmed with the authentic optical isomers.

REFERENCES

1. S. Ahuja, Discovery of New Compounds by Thin Layer Chromatography,
    Techniques and Applications of Thin Layer Chromatography, J. Touchstone
    and J. Sherma, Eds., Wiley, 1985, p. 109.
2. S. Ahuja, Ultratrace Analysis of Pharmaceuticals and Other Compounds of
    Interest, Wiley, N.Y., (1986).
3. B. Kratochvil, and J. K. Taylor, Anal. Chem., 53, 924A (1981).
4. W. J. Youden, J. Assoc. Off. Anal. Chem., 50, 1007 (1967).
5. F. W. Karasek, R. E. Clement, and J. A. Sweetman, Anal. Chem., 53, 1050A
    (1981).
6. S. Ahuja, P. Liu, and J. B. Smith, FACSS 9th Meeting, September 23, 1982.
7. S. Ahuja, Drug Quality Lecture, Indian Pharm. Congress, Dec. 28, 1974.
8. S. Ahuja, J. Chromatogr. Sci., 17, 168 (1979).
9. S. Ahuja, J. Pharm. Sci., 17, 163 (1976).
10. S. Ahuja, J. Liq. Chromatogr., 11, 2175 (1988).
11. J. A. Mollica, S. Ahuja and J. Cohen, J. Pharm. Sci 67, 443 (1978).
12. T. Yamana, and A. Tsuji, ibid, 65, 1563 (1976).
13. D. C. Chatterji, A. G. Frazier, and J. F. Galleli, ibid, 67, 622 (1978).
14. K. M. McErlane, N. M. Curran, and E. G. Lovering, ibid. 66, 1015 (1977).
15. S. Ahuja, P. Liu, and J. Smith, 45th International Congress of Pharmaceutical
    Sciences, Montreal, September 2-6, 1985.
16. S. Ahuja, J. Liq. Chromatogr., 11, 2175 (1988).
17. L. A. Svensson, Acta Pharm. Suec., 9, 141 (1972).
18. S. Ahuja, and C. Spitzer, Upublished Information, October 27, 1966.
19. S. Ahuja, D. Spiegel, and F. R. Brofazi, J. Pharm. Sci., 59, 417 (1970).
20. D. Spiegel, S. Ahuja, and F. R. Brofazi, ibid, 61, 1630 (1972).
21. S. Ahuja, Selectivity and Detectability Optimizations in HPLC, Wiley, N.Y.,
    1989.
22. R. M. Riggin, and C. C. Howard, Anal. Chem., 51, 210, 1979.
23. R. C. Weast, Handbook of Chemistry and Physics, C. R. C. Press, Inc., Boca
    Raton, p. C664, 1985.
24. S. Ahuja, G. Thompson, and J. Smith, Eastern Analytical Symposium, New
    York, September 13-18, 1987.
25. F. Matsui, E. G. Lovering, N. M. Curran, and J. R. Watson, J. Pharm. Sci.,
    72, 1223, 1983.

26.   S. Ahuja, S. Shiromani, G. Thompson, and J. Smith, Personal Communication, March 2, 1984.

27.   S. Ahuja, G. Thompson, and J. Smith 1st International Symposium on Separation of Chiral Molecules, Paris, France, May 31-June 2, 1988.

28.   S. Ahuja, Chiral Separations by Liquid Chromatography, ACS Symposium Series 471, American Chemical Society, Washington, D.C., 1991.

29.   G. Blaschke, H. P. Kraft, K. Fickentscher, and F. Koehler, Arzneim.-Forsch. 29, 1640 (1979).

30.   "Top 200 Drugs in 1982", Pharmacy Times, p. 25 (April 1982).

31.   Y. Okamoto, CHEMTECH. p. 176 (March, 1987).

32.   S. Ahuja, Analytical Profiles of Drug Substances, Vol. 4, p. 527, F. Florey, Ed., Academic, N.Y., 1985.

33.   H. R. Olpe, H. Demiéville, V. Baltzer, W. L. Bencze, W. P. Koella, P. Wolf and H. L. Haas, Eur. J. Pharmacol., 52, 133 (1978).

34.   R. P. Weatherby, R. D. Allan and G. A. Johnston, J. Neurosci. Method, 10, 23 (1984).

35.   D. F. Smith, and W. H. Pirkle, Phychopharmacol., 89, 392 (1986).

36.   E. W. Wuis, E. W. Beneken Kolmer, L. E. Van Beijsterveldt, R. C. Burgers, T. B. Vree and E. van der Kleyn, J. Chromatogr., 415, 419 (1987).

37.   H. Spahn, D. Krauβ, and E. Mutschler, Pharm. Res., 5, 107 (1988).

RECEIVED June 15, 1992

Chapter 3

Alternate Strategies for Analysis of Pharmaceuticals

John A. Adamovics and Kimberly Shields

Cytogen Corporation, Princeton, NJ 08540–5309

An increasingly popular chromatographic technique is to use unbonded silica gel column with aqueous mobile phases. This approach has been used to analyze acidic, basic and neutral pharmaceuticals in product formulations and biological samples.

HPLC on chemically bonded aqueous silica packings is the traditional choice for analysis of pharmaceutical dosage forms and clinical samples. Over the last 10 years there has been an increasing number of publications that have demonstrated the utility of using aqueous eluents with non-bonded silica for the analysis of a wide array of pharmaceuticals. Table I lists the published drug product assays that have used aqueous silica methods. Table II lists all the pharmaceuticals that have been analyzed in biological fluids.

These two tables represent over 120 different pharmaceuticals that have been analyzed by aqueous silica methods. Besides the pharmaceuticals listed in the two tables, several general studies have been published in the screening and quantification of basic drugs of forensic interest in biological fluids (34-40). The following sections describe the chromatographic parameters that affect selectivity and resolution of aqueous silica methods. The basic amino acid arginine, along with the compounds whose structures are shown in Figure 1, were studied.

Method Development

Changes in mobile phase composition such as the organic co-solvent, pH, ionic strength, and addition of surfactants have been systematically studied (1-3, 5, 7, 9, 12, 19, 22, 33, 35, 39). These studies demonstrate that retention of basic analytes is mediated primarily by the cation exchange properties of the silica. Retention of acidic and neutral analytes is not as clearly understood but could involve the siloxane moieties of silica, hydrogen bonding or other non-specific interactive forces.

TABLE I. Bulk and Formulated Drugs Analyzed by Aqueous Silica Methods

Drug	Pharmaceutical Preparation	Reference
Analgesics non-steroidal	Capsules and tablets	1, 2
Antihistamines, antitussives, and decongestants	Syrup and tablets	3, 4
Antihypertensives opthalmic solution and tablets	Transdermal patch	5, 6
Antibiotics and antimicrobial agents	Bulk and injectables	6-9
Aspartame	Bulk	10
Catecholamines	Various formulations	11
Famotidine	Oral dosages	12
Fosinopril	Bulk	13
Hydoxyzine hydrochloride	Syrups	14
Imidazoline derivatives and nasal drops	Capsules, ointments,	15
Pralidoxime chloride	Injectable	16
Prazosin hydrochloride	Capsules	17
Pseudoephedrine hydrochloride	Syrup and tablet	18
Steroids	Bulk	19
Succinylcholine	Injectable	20
Tripolidine hydrochloride	Syrup and tablets	21

TABLE II. Pharmaceuticals in Biological Fluids Analyzed by Aqueous Silica Methods

Drug	Biological Fluid	Reference
Albuterol	Plasma	22
Amiloride	Plasma and urine	23
Amiodarone	Plasma and serum	24
Anabolic Steroids	Serum	25
Antibiotics and antimicrobial agents	Plasma	9
Chlorpheniramine	Plasma and Urine	26
Eseroline	Plasma	27
Metoclopramide	Plasma	28
Morphine	Urine and blood	27, 30
Nitrazepam	Plasma	31
Physostigmine	Plasma	27
Tricyclic antidepressants	Plasma	32

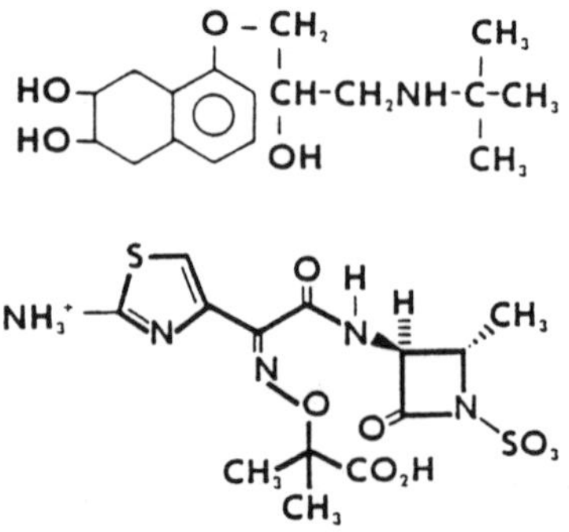

Figure 1.   Chemical structures of nadolol (top) and aztreonam (bottom).

Co-Solvent. The effect of the eluent water content upon retention of our test compounds at a constant acid composition of 0.1% is shown in Figure 2. Two of these compounds are the E and Z isomers of the antimicrobial aztreonam which is a monobactam containing a sulfonic acid moiety. These two isomers, which differ around the imino side-chain, are initially resolved with a mobile phase of 0.1% of phosphoric acid. When the eluent composition is greater than 20% acetonitrile, these isomers are no longer resolved. With additional quantities of acetonitrile there is a slight decrease in the retention of the monobactam. The amino acid arginine shows the greatest increase in retentiveness with increasing acetonitrile content while the β-adrenergic blocking agent, nadolol, shows a moderate increase relative to arginine.

This is in contrast to relatively non-polar molecules such as steroids where small changes in acetonitrile concentration lead to disproportionate change in retention time (25). For instance, a 10% acetonitrile concentration does not elute lypophilic steroids, while a 25% eluded them in the solvent front (25).

The substitution of methanol for acetonitrile results in very similar retention curves for arginine and nadolol. In contrast, as little as 5% methanol co-elutes the E/Z isomers of aztreonam. The substitution of methanol for acetonitrile for steroids requires twice the concentration to achieve similar retentiveness (25).

pH. At constant ionic strength, alterations in the eluent pH have the greatest impact on the retention of the basic compounds, arginine and nadolol as would be expected by a cation-exchange mechanism. Similar changes in selectivity are observed for the increase in eluent pH as were seen for increases in concentration of acetonitrile. The primary difference between change in the co-solvent versus pH is retentiveness which is of a greater magnitude with the pH alterations than with changes in co-solvent. There is an increase of the capacity factor with increase of eluent pH as the eluent pH approaches the pKa of the analyte where maximum retention for that analyte is obtained. This relationship has been extensively studied with other basic analytes (3, 33, 36).

Alteration in eluent pH not only affects the protonation of basic analytes, but also the ionization of the surface silanols (36). In order to unambiguously ascertain the influence of surface silanols, the effects of pH on the retention of several quaternary ammonium compounds have been studied (15, 36). With the increase in pH eluent (0-11) quaternary ammonium compounds increase in retention similar to the ionization profile of silica silanols. The retention behavior is attributed to the various types of silanols present on silica, the strongly acidic silanol sites are ionized at low pHs and the weakly acidic sites are ionized at neutral or basic pH's (15, 36).

Changes in pH eluent also affects the detector response of basic analytes. For instance, five alkylamines are resolved and detected using an acidic 0.5% phosphoric acid mobile phase with direct conductivity detection as shown in Figure 3. At higher pH's where the amines are no longer charged

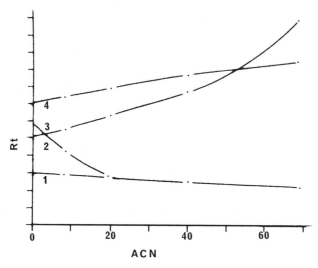

Figure 2.   Effect of increasing eluent concentration of acetonitrile on retention. Initial mobile phase composition was 0.1% phosphoric acid. Curves: 1 = Z-aztreonam, 2 = E-aztreonam, 3 = arginine, 4 = nadolol.

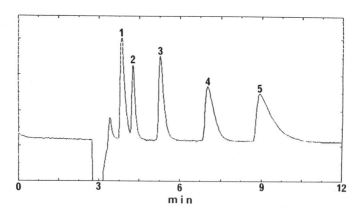

Figure 3.   Silica Chromatogram of 1. triethylamine; 2. 5-aminovaleric acid; 3. 6-aminocaproic acid; 4. trimethylamine; 5. choline. Mobile phase: 0.5% phosphoric acid with detection by conductivity (Waters).

except for choline, only choline will be detected.   Ultraviolet and amperometric detection of other amines has also been shown to be influenced by the eluent pH (36).

Another chromatographic characteristic that is influenced by pH is the peak shape. Factoring out the influence of retention, some drugs such as amphetamines give better peak shapes under strongly acidic conditions than at neutral pH, while drugs such as quinine show the reverse chromatographic characteristics. It is speculated that this may be due to differences in the solvation of the basic analytes at the various eluent pH's (36). Acidic analytes such as the isomers of aztreonam co-elute with eluent pH's above 3. At higher pH's, these co-eluting isomers show a further decrease in retention, but interestingly, are still retained on the silica column. This has also been observed for other acidic drugs such as aspirin, ibuprofen and naproxen (1).

Retention of neutral compounds such as steroids are not affected by changes in pH (25).

Ionic Strength. When the ionic strength of this eluent is increased by the addition of ammonium phosphate, the basic analytes show the most significant decreases in retention. Aztreonam shows only a slight decrease in retention as depicted in Figure 4. For neutral compounds no apparent changes in retention are observed (25). These observations are consistent with an ion-exchange mechanism where the k' values are strongly dependent on the concentration of cations in the eluent due to the competitive displacement from the ion exchange sites.

At neutral eluent pH's the increase in the ionic strength has shown to change the selectivity of some basic analytes. The change in selectivity is believed to be due to the presence of a portion of each base in the non-protonated forms, particularly at a pH near their pKa (35).

Surfactants. The addition of surfactants to the mobile phase has been cited as being an important variable in improving the column-to-column and lab-to-lab reproducibility of aqueous silica methods (41). The most commonly used additive has been cetyltrimethylammonium bromide (CTMA) which is a C-16 quarternary ammonium compound (5, 42-44). When using this quarternary salt, the silica is modified to create an, in situ, alkyl-bonded silica stationary phase (5).

As expected, non-ionic compounds are resolved mainly by hydrophobic interactions, anionic analytes are separated by a combination of hydrophobic and paired-ion effects. The retention of cations is based on ion exchange.

In the study of timolol, a β-adrenergic blocking agent which contains a secondary amine, the system selectivity is controlled by alterations in the CTMA and methanol concentrations. Ionic strength and/or pH variations give a less predictable retention. A combination of separation mechanisms appear to be at play; timolol follows an ionic exchange mechanism, while some of its degradants were more governed by reversed-phase mechanisms (5). Other dynamically modified silica systems have also been defined by using hydrazinium derivatives (45) and ethylenediamine based buffers (41). The disadvantage of these types of chromatographic systems are the long equilibration time (43).

Applications

Impurities-degradants. The impurities and degradants of bulk nadolol can be analyzed by either reversed-phase or aqueous silica methods (7). Table III compares the relative retentions of these impurities by these two chromatographic methods. When using the identical column dimensions, flow rate and detection, a run time of 25 minutes is required for nadolol and its first five impurities to elute from the reversed-phase column. Impurity 6 fails to elute under these conditions. Whereas on the silica system, impurity 6 elutes first. Nadolol and the remaining impurities elute in less than 10 minutes as shown in Figure 5.

Cephradine, a semi-synthetic cephalosporin which contains a cyclohexadiene moiety under stressed oxidative conditions degrades into cephalexin and m-hydroxycephalexin. Structures are shown in Figure 6 and a chromatogram of stressed cephredine spiked with p-hydroxycephalexin shown in Figure 7. In order to resolve m-hydroxycephalexin from p-hydroxycephalexin, the phosphoric acid content was adjusted from 0.1% to 0.05%.

Fosinopril, an angiotensin-converting enzyme inhibitor used to treat hypertension is resolved from its diastereoisomers and related impurities on silica. In contrast to the previous examples, the optimal eluent contains predominantly acetonitrile. The structure of fosinopril and its related impurities are shown in Figure 8. Figure 9 shows the resolution of fosinopril and its related impurities. The quantities of acid and water in the mobile phase are crucial in achieving this resolution and selectivity (13).

Drug Serum Analysis. Nearly all methods developed for the analysis of antibiotics in biological fluids are based on the use of reversed-phase chromatographic columns. Only under certain circumstances has it been possible to develop reversed-phase methods based totally on aqueous eluents. Otherwise, eluents with an organic modifier, especially greater than 15% over numerous injections will cause precipitation of the serum protein followed by obstruction and rapid deterioration of the chromatographic column. Consequently, the methodological approach analysis of antibiotics has been traditionally to deplete the sample of protein prior to chromatography. Besides the increased analysis time in using this approach, quantitative errors due to losses of drugs and proteins will occur.

Human serum can be directly injected on aqueous silica columns or when serum is not clear diluted one-to-one with eluent prior to injection (9). After about 15 injections of 20 microliters each, the chromatographic column becomes saturated with proteins and fats. Consequently, these components are no longer absorbed by the chromatographic column and elute in the void volume. The protein saturation of the column has no apparent detrimental affect on the quantifying of solutes as can be seen in Figure 10 where cephalexin and cephradine are resolved from control serum.

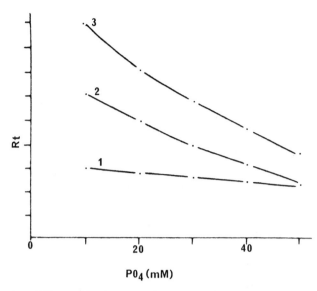

Figure 4.   Effect of ionic strength on retention.
Mobile phase: acetonitrile-water (50:50, pH 4)
Detection: UV 220nm. Curves: 1-aztreonam, 2-nadolol,
3-arginine.

TABLE III.   Relative Retention Times of Nadolol and Potential Impurities

Compound	ODS[a]	Silica[b] Gel
Nadolol	1.00	1.00
Compound 1	0.90	0.19
Compound 2	3.30	0.29
Compound 3	3.60	0.87
Compound 4	5.30	0.19
Compound 5	6.10	0.78
Compound 6	>10	0.15

[a] Mobile phase: 0.1M sodium perchlorate (0.15% perchloric acid)-
methanol (6:4)
[b] Mobile phase: 10mM ammonium phosphate (pH 7)-acetonitrile
(4:6)

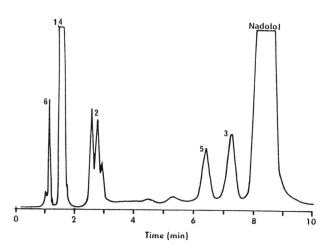

Figure 5.   Silica chromatogram of nadolol that has been spiked with several impurities.

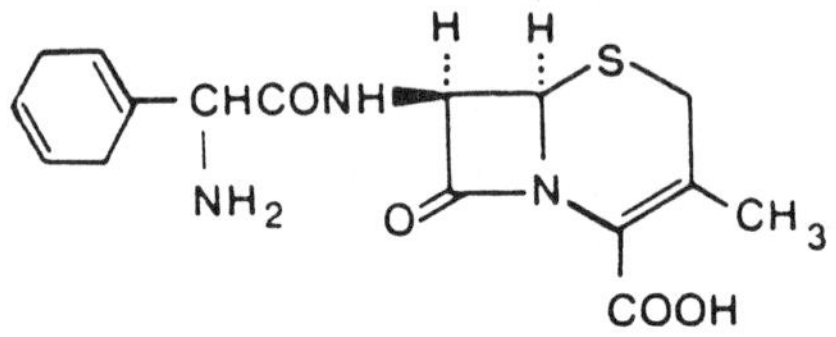

Cephradine

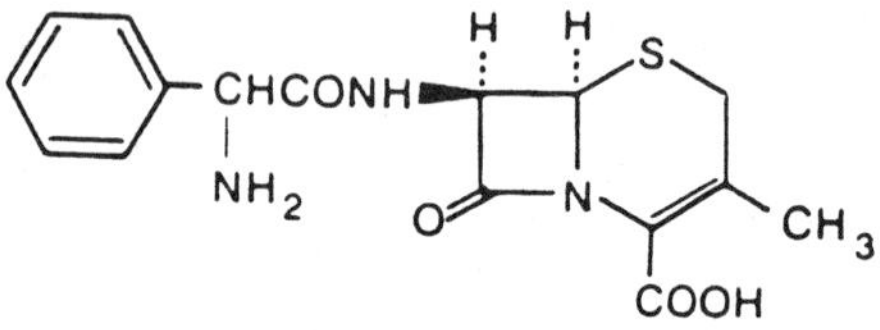

Cephalexin

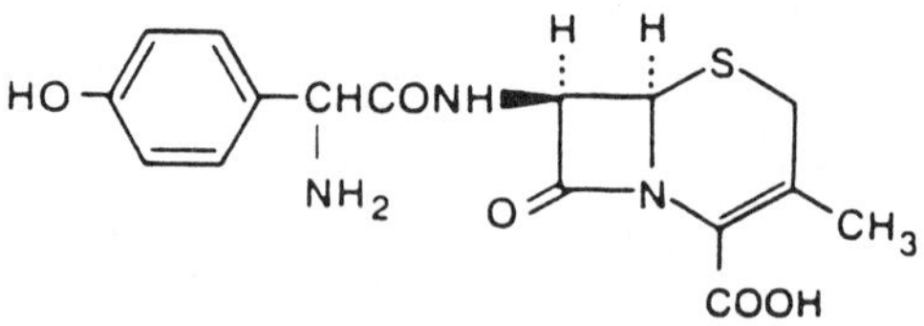

Cefadroxil

Figure 6.   Structures of cephradine and related cephalosporins.

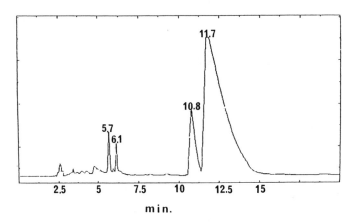

Figure 7.   Silica chromatogram of stressed bulk cephradine spiked with
p-hydroxycephalexin (cefadroxil) at 5.7 min,
m-hydroxycephalexin at 6.1 min, cephalexin at 10.8 min and
cephradine at 11.7 min. Mobile phase: 0.05% phosphoric
acid.

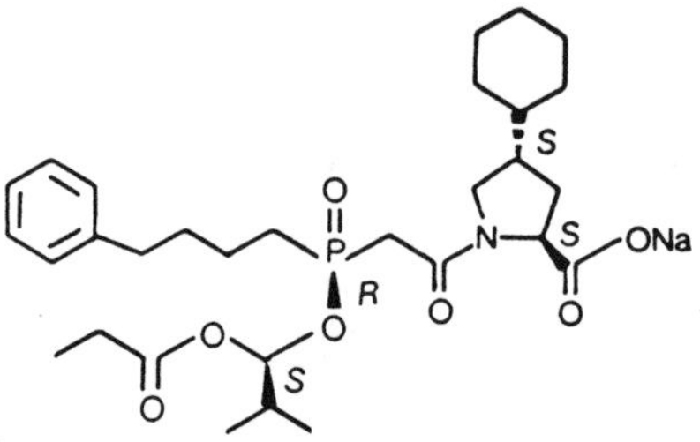

Fosinopril Sodium
(S,R,S,S)
3. (S,R,S,R)

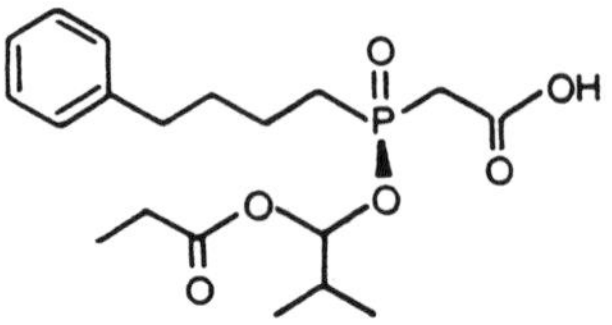

1. (R,S) and (S,R) acids

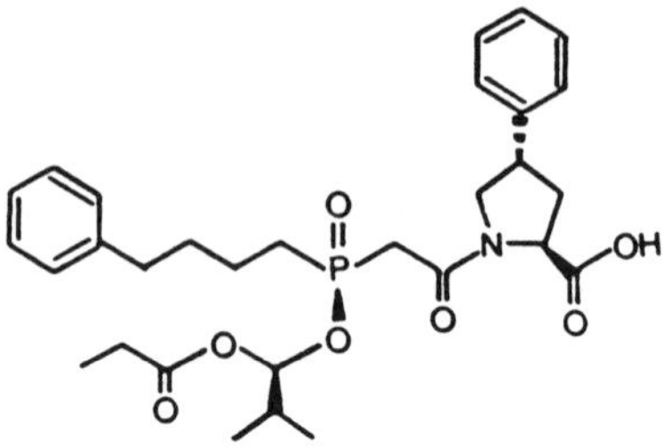

2

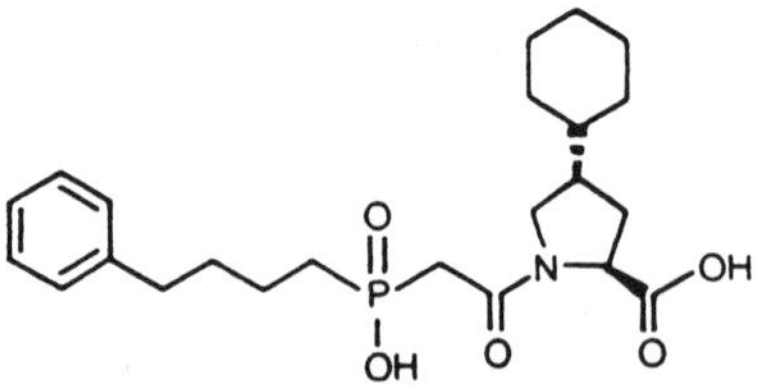

4

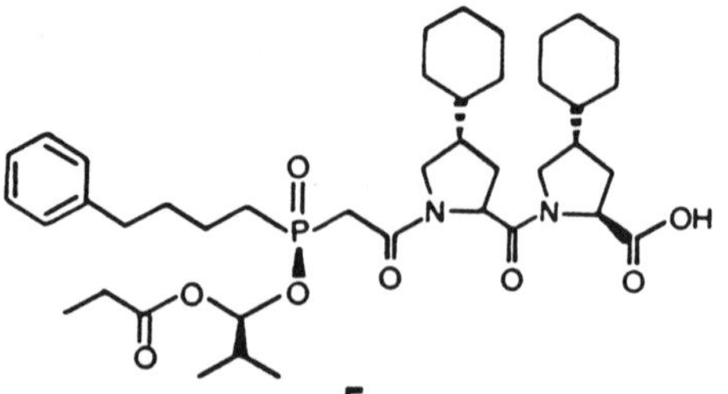

5

Figure 8. Structures of fosinopril and related impurities.

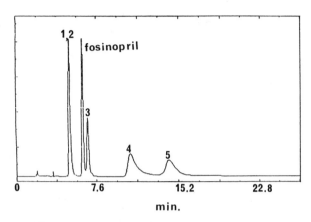

Figure 9.   Silica chromatogram of fosinopril and spiked impurities. The
mobile phase consists of acetonitrile-water-phosphoric acid
(4L:20 mL:2 mL) and detection at 214 nm.

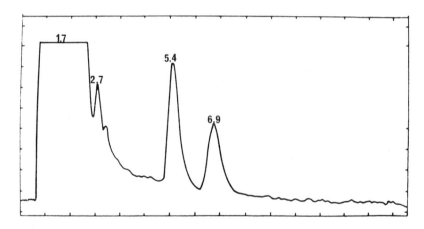

Figure 10. Chromatogram of cephalexin (5.4 min) and cephradine (6.9
min) spiked in control serum after 50 injections. Mobile
phase consisted of 0.5% aqueous phosphoric acid (9:1, V/V)
with monitoring at 220nm.

Conclusions

The utility of aqueous silica chromatographic methods has been demonstrated for a wide array of pharmaceuticals from bulk drug to biological fluids. The development of a chromatographic method for basic drugs is relatively straightforward and is based on the cation-exchange properties of silica and the pKa of the analyte. Acidic and neutral drugs are resolved by mechanisms that are not clearly understood, consequently, methods development requires an empirical approach to solvent optimization.

Literature Cited

1. Lampert, B. M.and Stewar, J. T. J. Chromatogr. 1990, 504, 381.
2. Kryger, S. and Helboe, P. J. Chromatogr. 1991, 539, 186.
3. Bildenmeyer, B. A.; Del Rios, J. K. and Korpi, J. Anal. Chem. 1982, 54, 442.
4. Law, B.; Gill, R. and Moffat, A. C. J. Chromatogr. 1984, 301, 165.
5. Mazzo, D. J. and Snyder, P. A. J. Chromatogr. 1988, 438, 85.
6. Adamovics, J. and Unger, S. J. Liq. Chromatogr. 1986, 9, 141.
7. Adamovics, J. LC, 1984, 2, 393.
8. Moats, W. A. and Leskinen, L. J. Chromatogr. 1987, 386, 79.
9. Adamovics, J. A. J. Pharm. Biomed. Anal. 1984, 2, 167.
10. Di Bussolo, J. M. and Miller, N. T. Seventh International Symposium on HPLC of Proteins, Peptides and Polynucleotides. 1987, Poster 108.
11. Helboe, P. J. Pharm. Biomed. Anal. 1985, 3, 293.
12. Buffar, S. E. and Mazzo, D. J. J. Chromatogr. 1986, 353, 243.
13. Kirschbaum, J.; Noroski, J; Cosey, A.; Mayo, D and Adamovics, J. J. Chromatogr. 1990, 507, 165.
14. The United States Pharmacopeia, Twenty-first Revision, United States Pharmacopeial Convention,Inc., Rockville, MD, 1985, 520.
15. Lingeman,H.; VanMunster, H. A.; Beyner, J. H.; Undersberg, W. J. M. and Hulshoff, A. J. Chromatogr. 1986, 352, 261.
16. Schroeder, A.C.; Di Giovanni, J.H.; Von Bredow, J. and Heiffer, M. H. J. Pharm. Sci. 1989, 78, 132.
17. The United State Pharmacopeia, Twenty-first Revision, United States Pharmacopeial Convention, Inc., Rockville, MD. 1985, p.869.
18. The United States Pharmacopeia, Twenty-first Revision, United States Pharmacopeial Convention, Inc., Rockville, MD. 1985, p.913.
19. Helboe, P. J. Chromatogr. 1986, 366, 191.
20. The United States Pharmacopeia, Twenty-first Revision, United States Pharmacopeial Convention,Inc., Rockville, MD. 1985, pp.986.
21. The United States Pharmacopeia, Twenty-first Revision, United States Pharmacopeial Convention,Inc., Rockville, MD. 1985, p.1099.
22. Wie, Y.Q.; Shi,R.; Williams, R.L and Lin, E. T. J. Liq. Chromatogr. 1991, 14, 253.
23. Shi, R. J. Y.; Bonet, L. Z. and Lin, E. T. J. Chromatogr. 1986, 377, 399.
24. Flanagan, R. J.; Stoney, G. C. A and Holt, D. W. J. Chromatogr. Sci., 1979, 16, 271.

25. Lampert, B.L. and Stewart, J. T. J. Liq. Chromatogr. 1989, 12, 3231.
26. Shi, R. J. Y.; Gee, W. L.; Williams, R. L. and Lin, E. T. J. Liq. Chromatogr., 1987, 10, 3101.
27. Wu, Y.Q.; Reinecke, E., Lin, E. T., Theoharidos, A. D. and Fleckenstein, L. J. Liq. Chromatogr., 1990, 13, 275.
28. Shi, R. J. Y.; Gee, W. L.; Williams, R. L. and Lin, E.T. Anal. Lett. 1987, 20, 131.
29. Jane, I. and Taylor, J. F. J. Chromatogr. 1975, 109, 37.
30. White, M. W. J. Chromatogr. 1979, 178, 229.
31. Kelly, H.; Huggett, A. and Dawling, S. Clin. Chem. 1982, 28, 1478.
32. Watson, I.D. and Stewart, M. J. J. Chromatogr., 1975, 110, 389.
33. Schmid, R. W. and Wolf, C. Chromatographia, 1987, 24, 713.
34. Jane, I. J. Chromatogr., 1975, 111, 227.
35. Flanagan, R. J. and Jane, I. J. Chromatogr., 1985, 323, 173.
36. Jane, I.; McKinnan and Flanagan, R. J. J. Chromatogr., 1985, 323, 191.
37. Greving, J.E.; Bouman, H.; Jonkman, J. H. G.; Westenberg, H. G. M. and DeZeeuw, P. A. J. Chromatogr., 1979, 186, 683.
38. Crommen, J. J. Chromatogr. 1979, 186, 705.
39. Kelly, M. T. and Smyth, M. R. Analyst, 1989, 114, 1377.
40. Kelly, M. T. and Smyth, M. R. J. Pharm. Biomed. Anal., 1989, 7, 1757.
41. Smith, R. M.; Westlake, J. P.; Gill, R and Osselton, D. M. J. Chromatogr., 1990, 514, 97.
42. Hansen, S. H.; Helboe, P. and Thomsen, M. J. Pharm. Biomed. Anal., 1984, 2, 165.

43. Hansen, S. H.; Helboe, P.; Thomsen, M. and Lund, U., J. Chromatogr., 1981, 210, 453.
44. Hansen, S. H. and Helboe, P. J. Chromatogr., 1984, 285, 53.
45. Shatz, V. D.; Sahartova, O. V. and Kalvins, I. J. Chromatogr., 1990, 521, 19.

RECEIVED June 3, 1992

Chapter 4

Ruggedness of Impurity Determinations in Pharmaceuticals

High-Performance Liquid Chromatography Analysis with Ultraviolet Detection

M. P. Newton, John Mascho, and Randy J. Maddux

Quality Assurance Department, Glaxo, Inc., Zebulon, NC 27597

The use of relative response factors is a good means to obtain weight percent impurity values when employing HPLC analysis with UV detection. The ruggedness of the relative response factor for a given reference and impurity compound is inversely proportional to the difference in shape and slope of the respective UV curves near the analytical wavelength. The values were found to be rugged for the chromatographic parameters studied, however it is recommended that UV detectors be calibrated for wavelength accuracy and the relative response factor values confirmed when changing detector models.

This is a preliminary investigation to study, from a semi-quantitative approach, the effects of several chromatographic/detector parameters on the ruggedness of impurity relative response factors.

HPLC is commonly used in the pharmaceutical industry to determine the purity profile of both drug substance and drug product samples. Once a compound and its related impurities are separated by the chromatographic process, there is considerable freedom as to the detection system to be used for quantitation purposes. The most commonly used detection system is UV spectroscopy which is acceptable for most compounds when using a variable wavelength detector. One of the major problems encountered when using UV detection is that the detector response can vary greatly from compound to compound at a given wavelength.

Once a separation has been developed and the detector wavelength selected, data can easily be generated in the

0097–6156/92/0512–0040$06.00/0

form of area or weight percents. It is usually most desirable to report impurity results in terms of weight percent since area percent data generated by a UV detector can be very misleading. One approach to obtaining weight percent data is to assay each impurity versus a standard of that same compound just as the major component is assayed. This is one accurate way to obtain weight percents, however it has several disadvantages. First, obtaining a reasonably pure standard of each compound, in sufficient quantity, may be a problem. Also, if a given compound is not stable, it will have to be prepared periodically and its purity redetermined. Standard solutions will have to be stored and solution stability will have to be determined. If any compounds are not stable in solution, separate solutions of these unstable compounds will have to be prepared frequently. All this increases the amount of preparation time, solvents used and analysis time. In addition the calculations could become quite cumbersome as the number of impurities increase. Finally, it is usually undesirable to have to provide other laboratories with standards of the impurities of known purity on a continuing basis.

One way to circumvent these problems is to generate relative response factors. A relative response factor is the ratio of the absorptivity of the impurity to the absorptivity of the main (reference) compound under the given analytical conditions. Another way of expressing this is as the ratio of the peak response per unit concentration for the impurity to the peak response per unit concentration for the reference compound. The response of the impurity in the sample solution can be compared to the response of the main compound in the external reference standard solution for which the concentration is known. If the main compound is close to 100 percent pure in a drug substance or near to the theoretical amount in a drug product, then the response of the main compound in the sample may be used as a reference. The area percent of the impurity is then simply divided by its relative response factor to obtain the weight percent.

Some advantages of using relative response factors is that impurity standards, with known purities, have to be prepared only once. A relatively small quantity is required and the impurity standard has to be chromatographed only on one occasion. When calculating relative response factors, it may be necessary to mathematically convert the concentrations of the impurity and reference compound standards used to that of the same solvate and salt form which would be present in the sample to be analyzed. Once the relative response factors have been determined for a series of impurities, a computer can be used to easily and automatically apply the proper relative response factor to each compound based on its retention time or relative retention time.

The accuracy of weight percent data is influenced by the ruggedness of the relative response factor value. Any parameter variation that could impact the relative absorptivity of the impurity compared to the reference compound under given analytical conditions could affect the relative response factor. In this investigation several parameters were varied and the effects on the relative response factors were studied.

Experimental

HPLC grade methanol and water and reagent grade ammonium acetate were used to prepare the mobile phase. Reagent grade sodium hydroxide and glacial acetic acid were used for the pH adjustment experiments. The standard conditions included a 95:5 methanol: 0.1M ammonium acetate (pH 7.6) mobile phase, 2 mL/min flow rate, 15 uL injection volume, 30°C column (and detector) temperature and detection at 228, 245, and 310 nm. One injection was made at each new condition. The column was allowed to equilibrate for thirty minutes at each new condition before the "spiked sample" was injected onto the column. The four compounds used throughout the study were dissolved in the mobile phase and all present in the same spiked sample solution. The reference compound and three impurities were at concentrations of about 5 times the limit of quantitation which was well within the linear range for each compound.

The equipment utilized included a 4.6 mm x 25 cm Spherisorb ODS1 (Keystone) column, Waters Model 710 pump, Hewlett Packard 1040 diode array detector, Waters Model 712 WISP autoinjector and either a Hewlett Packard Chem Station or a Hewlett Packard 3350 computer.

Response factors (peak response per unit concentration) were generated for one "reference" and three "impurity" compounds under standard and modified conditions. Relative response factors were then calculated by dividing the impurity response factors by the reference compound response factor, under identical conditions, i.e., obtained from the same chromatogram. The main numerical tool utilized in this study is the percent change in the relative response factor when going from standard conditions to a condition in which one parameter is changed. The relative response factor of an impurity at the standard condition is subtracted from the relative response factor obtained at the modified condition. That difference is divided by the relative response factor at the standard condition and then multiplied by 100.

For example, the concentrations of impurity 3 and the reference compound were 0.446 and 0.519 mg/mL respectively. The areas generated at 245 nm by impurity 3 at standard (pH 7.6) and pH 8.6 conditions were 378 and 348, respectively. The corresponding areas for the reference compound were 519 and 524. The relative response factors for impurity 3 at 245 nm under standard and pH 8.6 conditions are

$$\frac{378/0.446}{519/0.519} = 0.848 \quad \text{and} \quad \frac{348/0.446}{524/0.519} = 0.773, \text{ respectively.}$$

The change in the relative response factor for impurity 3 going from pH 7.6 to pH 8.6 is

$$\frac{(0.773 - 0.848)}{0.848} \times 100 = -9\%.$$

In Tables I and II changes in the relative response factors not exceeding $\pm$ 5% were left out so the more significant changes and therefore trends would not be obscured by "insignificant" or random differences.

Results

A chromatogram of the spiked sample solution obtained under standard conditions is shown in Figure 1. The four compounds were well separated under all conditions except for impurities 1 and 2 which were not resolved when the pH of the mobile phase was raised from 7.6 to 8.6.

The UV spectra of the four compounds studied are provided in Figure 2. The slope of the UV curves at each of the three wavelengths employed for this study (228, 245, and 310 nm) were estimated for each of the four compounds (Figures 3 - 6). The difference in slope of each impurity compared to the reference compound was as high as 120° and is provided in the data tables. The similarity of the UV curve shape between the reference compound and the impurities near the analytical wavelengths is also provided in the data tables.

The most critical parameter studied was the analytical wavelength. Varying the wavelengths -5 to +5 nm resulted in changes in the relative response factors of up to 84% (Table I). Other parameters expected to possibly change the UV characteristics of the four compounds are pH, percent organic composition (solvent strength), and ionic strength of the mobile phase. Changes in relative response factors up to 28, 15 and 13% respectively were observed when these parameters were modified (Table II). The reference compound used for this study was known to have a temperature dependent UV spectrum. Changes up to 12% in the relative response factors resulted when the temperature was increased by 20°C (Table II). As expected, two variables that did not result in any significant change in relative response factors were injection volume and flow rate (Table II). Five different models of variable wavelength UV detectors were used to evaluate the same spiked sample solution under standard conditions. The Hewlett Packard 1040 diode array detector was used as the reference. The

Table I. Percent Change in Relative Response Factors Due to Changes in Analytical Wavelength

Wavelength (nm)	228			245			310		
Compound #	1	2	3	1	2	3	1	2	3
ΔShape	D	S	I	D	S	S	D	D	S
ΔSlope	-60	-30	+20	+50	0	0	-120	-115	0
ΔWavelength									
+5nm	-13	-	+13	+33	-	-	-35	-33	-
+2nm	-10	-	-	+11	-	-	-16	-14	-
-2nm	-	-	-	-10	-	-	+16	+16	-
-5nm	+17	+6	-7	-17	+6	-10	+49	+44	-
Range	30	6	20	50	6	10	84	77	<5

D = Different
S = Similar
I = Intermediate

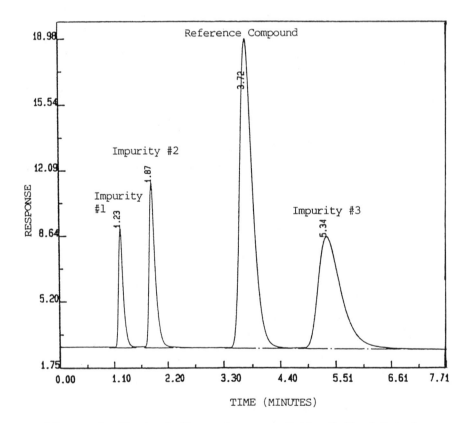

Figure 1. Typical Chromatogram of the Spiked Sample Obtained Under Standard Conditions

Table II. Percent Change in Relative Response Factors Due to Changes in pH, Solvent Strength, Ionic Strength, Detector Temperature, Injection Volume and Flow Rate

Wavelength (nm)	228			245			310		
Compound #	1	2	3	1	2	3	1	2	3
ΔShape	D	S	I	D	S	S	D	D	S
ΔSlope	-60	-30	+20	+50	0	0	-120	-115	0
ΔpH +1	(c) +21	(c) +8	-7	(c) +26	(c) +8	-9	(c) +28	(c) +23	-10
ΔpH -1	-	-	-	-7	-	-	-	-	-
Δ%MeOH +3 (a)	-6	-	-	-	-	-	-	-	+8
Δ%MeOH -3 (b)	+15	-	-	-	-	-	-	-	-
ΔIonic Strength +50%	+13	+7	-	-	-	-	-	-	-
ΔIonic Strength -50%	+8	-	-	-	-	-	-	-	-
Δ°C +10	-7	-	-	-11	-	-	-6	-	-
Δ°C +20	-10	-	-	-	-	-	-12	-6	-
ΔInjection Volume (uL) +7	-	-	-	-	-	-	-	-	-
ΔInjection Volume (uL) -7	-	-	-	-	-	-	-	-	-
ΔFlow rate +0.5 (mL/min)	-	-	-	-	-	-	-	-	-
ΔFlow rate -0.5 (mL/min)	-	-	-	-	-	-	-	-	-

(a) 92:8 MeOH: 0.1 $\underline{M}$ NH_4HOAC
(b) 98:2 MeOH: 0.1 $\underline{M}$ NH_4HOAC
(c) Impurities 1 and 2 were not resolved at pH 8.6

D = Different
S = Similar
I = Intermediate

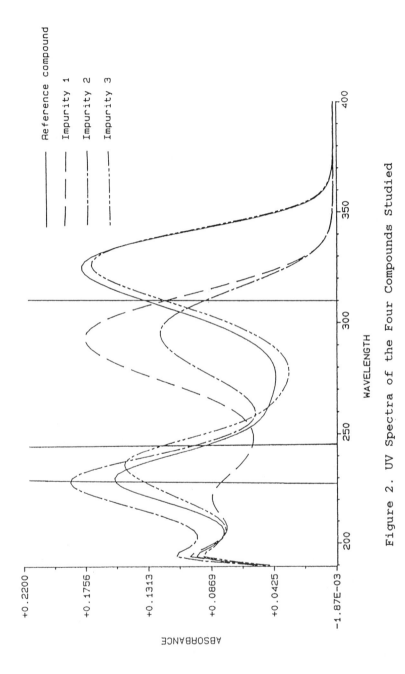

Figure 2. UV Spectra of the Four Compounds Studied

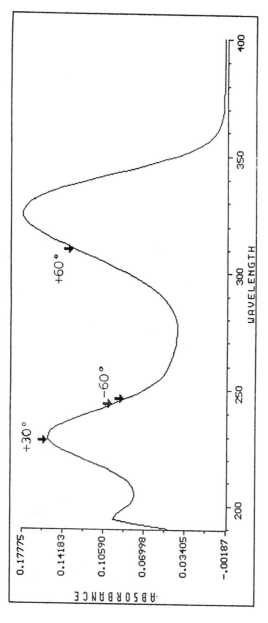

Figure 3. UV Spectrum of the Reference Compound

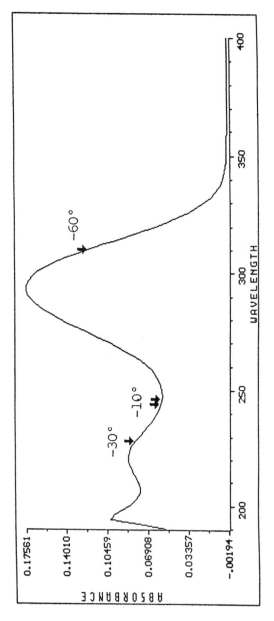

Figure 4. UV Spectrum of Impurity 1

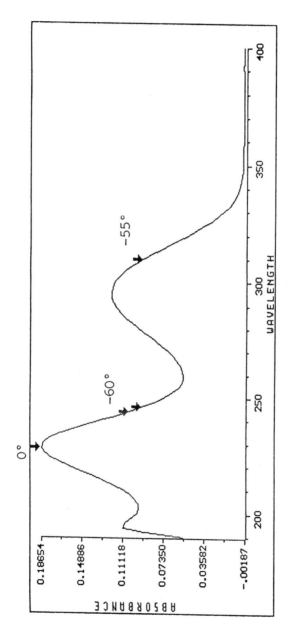

Figure 5. UV Spectrum of Impurity 2

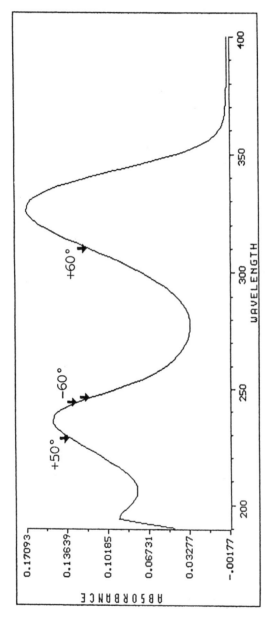

Figure 6. UV Spectrum of Impurity 3

relative response factors varied as much as 40% when different detectors were used (Table III).

Discussion and Conclusion

The conclusion drawn from the data is that when a chromatographic parameter is modified, the magnitude of the change in the relative response factor for a given reference and impurity compound is inversely proportional to the difference in slope and shape of the respective UV curves near the analytical wavelength. It should be noted that the wavelengths employed for this study were purposely chosen to maximize the unruggedness of the relative response factors in order to present a worse case scenario. When the wavelength was changed up to $\pm$ 5 nm the change in relative response factors ranged from < 5% to 20% in areas of small differences of the reference and impurity UV curves. The relative response factor change ranged from 30% to 50% for intermediate differences in the UV curves and 77% to 84% when large differences in the UV curves were present near the analytical wavelength (Table I).

This trend was also present when less sensitive parameters were modified i.e., pH, solvent strength and ionic strength of the mobile phase and detector temperature (Table II). However, even at wavelengths where the UV curves of the reference and impurity compounds varied greatly, when these parameters were changed more than would be expected in normal day to day operation, the relative response factors proved to be quite rugged. The majority of changes in the relative response factors were not greater than 5%. It should also be noted that for each compound, modifications of different parameters may result in some portions of the UV curves being affected more than others.

A parameter expected to have a significant effect on UV spectra, pH, in fact produced changes in relative response factors which ranged from <5% to 28%. The most significant changes were observed at the wavelengths where the slopes and shapes of the UV curves of the reference and impurity compounds exhibited the greatest differences. This same trend continued when the solvent strength, ionic strength and temperature were modified producing changes in relative response factors up to 15%. Modifying parameters not expected to have an influence on the UV spectrum, e.g., injection volume and flow rate, did not produce significant changes in the relative response factors (Table II).

Significant differences in relative response factors were sometimes observed when different UV detectors were used to analyze the same spiked standard solution. The range of differences between detectors was again proportional to the differences in slope and shape of the impurity and reference compound UV curves near the

Table III. Percent Change in Relative Response
Factors Due to Change in Detectors[a]

Wavelength (nm)	228			245			310		
Compound #	1	2	3	1	2	3	1	2	3
ΔShape	D	S	I	D	S	S	D	D	S
ΔSlope	-60	-30	+20	+50	0	0	-120	-115	0
Detector									
ABI 781	-19	-17	+4	-16	-17	+4	+3	+1	+4
Waters 484	-15	-19	+26	+9	-15	-2	-37	-36	+8
Waters 481	-14	-14	+6	-23	-16	+1	+1	0	+3
HP 1050	-22	-16	+8	-8	-18	+5	-21	-22	+4
Range	22	19	26	32	18	7	40	37	8

D = Different
S = Similar
I = Intermediate

[a]The HP 1040 detector was used as the reference.

analytical wavelength. At wavelengths where there were slight differences between the slope and shape of the UV curves, the average change in the relative response factor was 11%. The average change in relative response factors for intermediate slope and shape differences was 25% while an average change of 38% was observed for large differences. The average difference in relative response factors calculated for the three impurities at the three wavelengths studied ranged from 3% between the ABI 781 and Waters 481 detectors to 19% between the HP 1040 and Waters 484 detectors.

If relative response factors are to be used to quantitate impurity levels, the wavelength accuracy of the UV detector should be checked periodically and adjusted if not within a reasonable tolerance, e.g. $\pm$ 2 nm. Caution should be exercised when using different model detectors. A standard impurity solution could be useful in verifying or obtaining relative response factors for a detector model not used in the initial development of the method.

When developing HPLC methods for quantitating impurities, maximizing the ruggedness of relative response factors should be considered along with sensitivity and linearity requirements when selecting the analytical wavelength. The slope and shape of the UV curves of the compounds at the analytical wavelength should be as similar as possible. In addition, the chromatographic conditions chosen to optimize the separation and run time should be examined to optimize not only the ruggedness of the separation but also that of the relative response factors. Guidance should be provided in the analytical method as to how closely these parameters should be controlled to ensure assay integrity. It may be advantageous to monitor more than one wavelength in order to maximize the ruggedness of relative response factors when several impurities require quantitation. Many HPLC detectors are currently available that allow several wavelengths to be monitored simultaneously.

In conclusion, the use of relative response factors is a good means to obtain weight percent impurity values. To optimize the ruggedness of these values, UV curves of the reference compound and the impurities should be obtained in the mobile phase and be considered when selecting the analytical wavelength as discussed above. Other chromatographic parameters should also be evaluated to determine their influence on these relative response factors. The UV detectors used to quantitate impurities by way of relative response factors should also have their wavelengths calibrated periodically. Relative response factors generated with one model detector should be verified when changing to a different model detector whenever possible.

RECEIVED April 13, 1992

Chapter 5

High-Performance Liquid Chromatography in Pharmaceutical Development of Antibiotic Products

L. J. Lorenz

Lilly Research Laboratories, Eli Lilly and Company, Lilly Corporate Center, Indianapolis, IN 46285

The analytical properties and the preparative abilities of HPLC make it a very important tool in the development of antibiotic products. The drug development process requires evaluation of drug stability and of the types of degradation that may arise in the product. Both isocratic and gradient screening techniques are routinely used for these types of studies. HPLC and hyphenated HPLC techniques offer many handles for this type of work. HPLC, coupled with mass spectrometry or diode array detectors, can often offer some insight into the nature of the degradation product or impurity. Unfortunately, NMR information is usually required to definitively assign the structure of the compound. Preparative HPLC techniques offer the capability for obtaining suitable samples of degradation products for more sophisticated structural studies.

HPLC is an invaluable tool in the development of antibiotic products. HPLC has become the technique of choice for determining the potency and dose uniformity of a drug product or the purity of a bulk drug raw material. This technique also allows a chemist to establish the factor composition of a multifactored product. In addition, HPLC offers the ability to isolate and monitor impurities and degradation products. Once isolated, these compounds can be studied with modern spectroscopic techniques to establish structure and confirm the entity's identity.

Cefaclor, shown in Figure 1, is an important ß-lactam antibiotic that has undergone extensive development effort. Cefaclor capsules and tablets contain cefaclor as a dry crystalline powder in a monohydrate form. Cefaclor suspensions contain cefaclor as a dry crystalline powder until the product is constituted and dispensed to the patient. While the hydration of the molecule affects the stability and solubility properties of cefaclor, the actual solid form of cefaclor has little or no impact on the HPLC results.

Antibiotics are often labile compounds. This stability issue often dictates using special or very conservative sample handling procedures rather than those used for common stable pharmaceutical substances. Elaborate refrigerated sampling equipment and automated robotic sample processing are commonly used to accommodate these special needs for antibiotic substance evaluation.

0097–6156/92/0512–0054$06.00/0

Assay and Evaluation of Antibiotic Products

Simple, rugged assays are desired for the assay and evaluation of antibiotic products. Reversed phase HPLC systems are preferred for such applications. Most ß-lactam antibiotics can be chromatographed using several different separation mechanisms. Figure 2 shows a typical HPLC chromatogram for cefaclor. This assay is run in an acidic phosphate system containing pentanesulfonic acid as an ion pair reagent. Small amounts of triethylamine are also added to improve and maintain good peak shape over extended column use in the assay procedure.

Bulk Drug and Formulated Products. This same chromatographic system is used to assay both bulk drug materials and formulated products. The assays are primarily designed as a "dilute-and-shoot" type of application. The drug substance is dissolved in an aqueous acidic buffer containing an internal standard that provides suitable drug solubility and reasonable solution stability for the drug substance. The resulting solution may then be directly injected onto the chromatographic system with or without the use of a membrane filtration step before the injection. Formulated products are handled in a similar manner to provide the simplest preparation procedure possible.

Suspensions are handled by taking an aliquot of the suspension. For best results, the aliquots are weighed and converted to a volumetric with incorporation of the measured sample density. This procedure allows a dose per volume result to be calculated and eliminates problems associated with dispersed air bubbles in the suspension. When dealing with capsule or tablet products, the assay sample may be prepared either from a composite sample or by an individual dose approach. Procedures for suspensions or capsules and tablets involve a dissolution process that may incorporate a secondary dilution to provide an assay solution containing the appropriate concentration of the analyte and internal standard. After preparation of the assay solution, the solution is passed through a membrane filter and injected onto a chromatographic system. If the samples are prepared in a batch mode, the assay solutions are usually queued in a refrigerated autoinjector.

Internal Standard vs. External Standard. The cefaclor assay described here incorporates an internal standard that accounts for the second large peak in the assay chromatogram. Experiences in our laboratories have shown that the actual assay precision is usually slightly better when using an external standard approach.

There are still several good reasons why internal standards are routinely incorporated into the assay procedure. Several of these reasons are listed in Figure 3. One major reasons stems from the fact that our laboratory develops procedures that will be used in many different facilities in several different countries. Internal standards are still preferred in many foreign countries and in the laboratories of our foreign affiliates. Also, there are technical reasons for incorporating the internal standard into an assay procedure. The resolution between peaks is a useful monitor to show that the method performs properly throughout the sample run. Additionally, internal standards aid in deciding if an instrument malfunction occurred during the run and which data should be suspect.

With an external standard approach, rejection of an assay result can be difficult. The rejection of an individual piece of data from a large run requires the assignment of a reason for the deletion of the information from the data set. This is especially a problem when a very small bubble may have been present in the injected sample causing a low result. With the internal standard approach, it is easy to see that both peaks are low. In many situations, the internal standard corrects for the problem. In other cases, the internal standards provide a mechanism for rejecting an individual

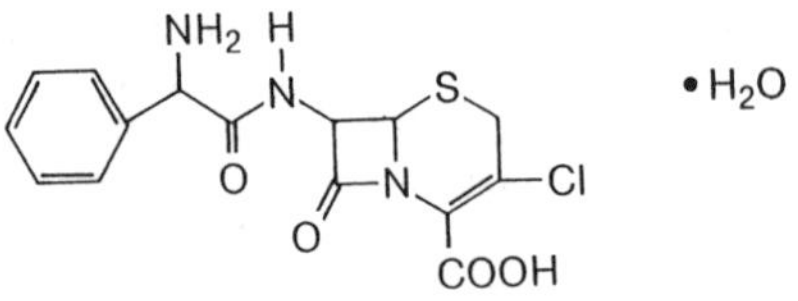

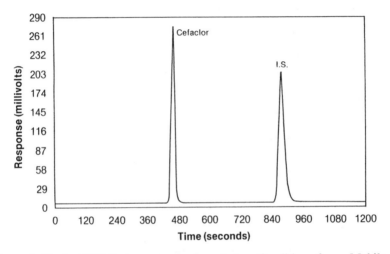

Cefaclor

Figure 1. Cefaclor structure.

Figure 2. Typical HPLC chromatogram for cefaclor. Condition of run: Mobile phase: 1 gram pentane sulfonic acid sodium salt, 780 mL water, 10 mL triethylamine. Adjust the pH to 2.5 with phosphoric acid. Add 220 mL of methanol. Column: Beckman Ultrasphere   ODS (25 cm). Flow rate: 1 mL/min. Detection: Ultraviolet at 265 nm.

- Laboratory preference
- Verification of system performance
- Verification of injection performance
- Provide justification for data rejection
- Verify process of automated sample processors

Figure 3. Justification for internal standards.

injection. If an injection problem occurs and both peaks in the chromatogram are low, there is a legitimate reason for rejecting the data from the assay average.

Robotic Sample Processors. The advent of robotic sample processors is yet another compelling reason for having internal standards present as a running check of the assay performance. Here the internal standard provides a valuable piece of information to validate or to ensure that the sample preparation and injection were properly performed. Even though internal standards have fallen somewhat into disfavor for sample quantitation in HPLC, one can see a resurgence in the need for internal standards in assay systems for validation purposes. Internal standards provide a good monitor of system performance and measurement criteria that aid in decisions about data acceptability. This need for an internal standard becomes even more critical when dealing with automated sample processing systems that will become the norm for performing assays in the future.

Determination of Impurities and Degradation Products

A very important part of antibiotic development programs is the determination of impurities, synthetic precursors, and degradation products for the drug substance. Again, researchers rely heavily upon HPLC techniques for these determinations. Sophisticated gradient HPLC procedures are frequently used during compound development efforts. Figure 4 shows a typical gradient chromatogram for cefaclor. These gradient procedures allow a sample to be monitored over a broad range of polarities to separate the different types of materials that may be present in a sample. Synthetic precursors and some degradation products are often much less polar than the drug entity itself, leading to the presence of late eluters in the chromatographic system. Additionally, many degradation products are much more polar than the drug itself, thus this set of polar compounds elutes early in the chromatogram. The examination of the entire range of potential sample components requires gradient HPLC.

Detection of Extraneous Substances. Selecting the means to detect extraneous substances is an important part of HPLC assay development. Ideally, diode array detectors can be used through some of the early developmental phases. These detectors can monitor many different wavelengths and can potentially detect many different compounds. Figure 5 is a typical chromatogram that can be obtained from this type of program. In this example, several compounds can be observed at lower wavelengths. Typically, the ultraviolet absorption maximum for cephalosporins lies near 260 nm. In this example, there are also many sample entities that are nondetectable at or near 260 nm. Therefore, monitoring at a lower wavelength provides more information than at higher wavelengths. Figure 6 shows an actual chromatogram monitored at 220 nm.

The chromatography system chosen for this work is a buffered system with acetonitrile serving as the modifier for the gradient. In this application for cefaclor, the separation of several of the major degradation products is very pH sensitive. This is shown in Figure 7, which is the pH retention profiles for several substances related to cefaclor. Several compounds cross and undergo reversal of retention as a function of change of pH. Therefore, good pH control is essential to maintain acceptable assay performance.

Identification and Quantitation of Impurities. Often, a major problem in drug development is identifying and quantitating impurities in a drug substance. Again, HPLC plays a very important role in this phase of drug development. In this phase of the drug development program, the drug product is stressed using heat, light, peroxide, acid, and base to establish the most likely degradation compounds that

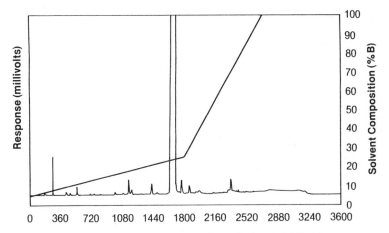

Figure 4. Typical gradient chromatogram for cefaclor. Mobile phase:
A solvent, 6.9 g/ L sodium phosphate monobasic adjusted to pH of 4 with
phosphoric acid. B solvent, 550 mL of solvent A with 450 mL of acetonitrile.
Gradient: Start at 5% of solvent B with a linear program increasing to 25% at a
rate of 0.67% change/ min. The second step programs to 100% B solvent at a
rate of change of 5% change/min. Flow rate: 1 mL/ min. Detection: ultraviolet
at 220 nm.

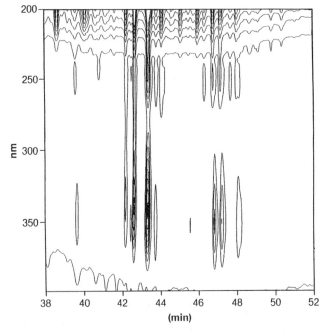

Figure 5. Typical HPLC diode array chromatogram. (See Figure 4 for
conditions.)

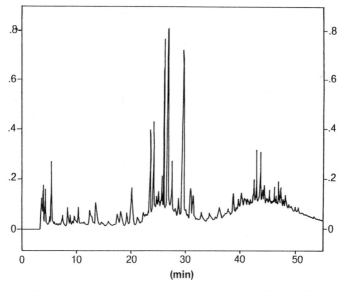

Figure 6. Chromatogram for cephalosporin monitored at 220 nm (See Figure 4 for conditions.)

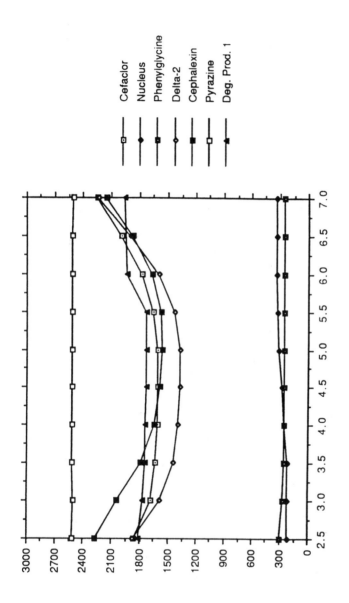

Figure 7. pH retention profiles for several cefaclor related substances. (Conditions are the same as given in Figure 4 except for the pH of the mobile phase. The pH was adjusted with phosphoric acid or sodium hydroxide to provide the appropriate pH.)

might form. When possible, these stress-generated compounds are compared to products that form during the actual storage life of the drug substance or product. Antibiotics are usually very sensitive to these types of conditions and form many different products. The detection of these compounds, again often with diode array detection devices, offers a way to identify and quantitate the major products that are formed under each of the conditions. These studies also serve as a basis for the selection of preparative HPLC systems for processing these samples.

Preparative HPLC Systems. During the development of rugged assays, one normally searches for good buffering agents to use in the reverse phase chromatographic systems. These buffers are usually selected to be ultraviolet-transparent to allow maximum flexibility in the selection of monitoring ultraviolet wavelengths. Typically, inorganic salts are selected as the normal HPLC buffer systems. However, salt-free systems are desired in preparative applications. Therefore, systems utilizing acetic acid, trifluoroacetic acid, and triethylamine are frequently used in the developmental phase that concentrates on the isolation and characterization of impurities. These substances can normally be removed by lyophilization techniques as part of the sample work-up procedure.

These criteria for preparative applications often dictate the development of additional systems for isolation and characterization work. Hyphenated systems such as HPLC-diode array systems and HPLC-mass spectrometry systems may also be used in this phase of the developmental program. Often these systems have unique requirements about buffers and additives present in the HPLC mobile phase.

Figure 8 is a chromatogram of the degradation products formed when cefaclor is exposed to heat and acid for the heat-induced and acid-induced degradation of cefaclor. Hyphenated techniques such as the HPLC-diode array device and HPLC-MS can often provide valuable information about sample impurities and degradents. Figure 9 is an HPLC diode array chromatogram of a portion of the cefaclor chromatogram showing the UV spectra for the individual components in the sample. This information helps to determine which compounds might be structurally related. This is shown vividly by the late eluters in this chromatogram where there are a whole series of degradents having similar spectra which are different from the spectrum of the parent drug substance. The addition of mass spectral data with this information can be an important part of the structural elucidation for such compounds.

Identification of Unknown Entities. These techniques might be sufficient to positively identify a known compound in the sample. Unfortunately, this information is not always sufficient to establish the identity of an unknown entity. Thus, the preparative features of HPLC must usually be applied to generate authentic isolated impurities that can be examined by additional techniques to allow the assignment of structure.

Semi-preparative techniques using columns up to two inches in diameter and sometimes larger are used in antibiotic development programs. Volatile buffers can be used in a gradient or isocratic mode. Several major chromatographic cuts are obtained from the degraded drug material. These initial chromatographic cuts are then lyophilized and rechromatographed one or more times until a single component entity is obtained. Following verification that one has a chromatographically pure entity, the material is submitted for sophisticated spectral characterization.

The determination of impurities and degradation products in an antibiotic drug such as cefaclor can become a very complex issue for the analytical chemist. One often must quantitate many materials. Some of the impurities are structurally similar to the drug substance. These entities will have spectral response characteristics similar to those of the parent drug material. For these types of substances, quantitation can be done versus the response of the parent drug substance. Another set of materials may

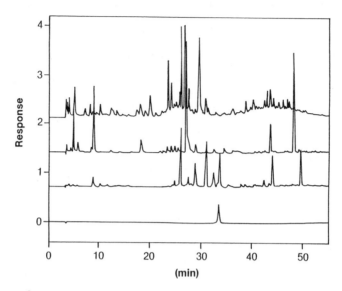

Figure 8. Chromatogram of heat -induced and acid-induced degradents of cefaclor. (Conditions are equivalent to those provided in Figure 4.)

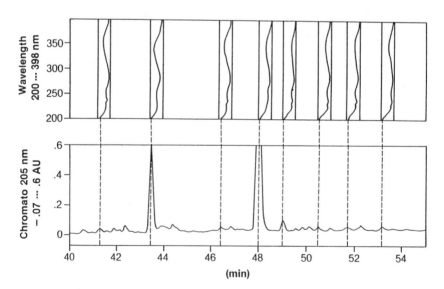

Figure 9.   HPLC diode array chromatogram for cefaclor showing the UV spectra for individual sample components. (See Figure 4 for conditions.)

be very different structurally and may have very different spectral absorption properties at the monitoring wavelength. One approach for determining impurities quantitates these compounds against authentic materials of the compounds in question. A second approach uses a response factor that is developed for the impurity versus some other compound. Frequently the response of the impurity is compared with the drug substance so that the drug substance can serve as a standard in the normal assay for the impurity.

Challenges for the Developmental Chemist

The actual chromatographic system for impurity analysis presents many challenges for the development chemist. Figure 10 lists some of the factors that must be considered. The chemist must consider the ease of use and ruggedness of the procedure. The method must be transferable and perform well in routine use in a production laboratory. The regulatory acceptability of the procedure must also be considered because they will need to react to procedures that will be put into official monographs, which often serve as the official control procedures for a drug. Many of the worldwide pharmacopeial associations tend to be very conservative about adopting sophisticated HPLC procedures. Today, the chemist must also consider how the procedures might be easily adapted to standard automated or robotic systems that process samples for chromatographic applications.

Automation. When dealing with the analysis of impurities in antibiotics such as cefaclor, the chemist frequently observes rapid changes in the actual impurity profile of the sample after sample dissolution. Thus, the sample must be prepared and injected onto the HPLC almost immediately after preparation. Since these assays often require 30 to 60 minutes for each injection, automation is required to allow for any reasonable throughput of samples. We have developed "dilute-and-shoot" robotic systems to support this type of operation. Such a robotic system is shown in Figure 11. This system queues samples in a dry state, adds diluent, dissolves the sample, filters the sample, and delivers it to the chromatographic system upon demand. This type of automation is very reliable. In fact, the reliability of the processing part of these types of systems is much better than the reliability of the chromatographic equipment. The limiting factor in these types of systems has become the reliability of the HPLC columns, pumps, and detectors associated with the evaluation system.

Gradient vs. Isocratic Approaches The chemist often must develop two different assay procedures for determining impurities and degradation: gradient reversed phase HPLC and isocratic systems. Gradient reversed phase HPLC procedures are very efficient in providing solutions to these problems. Figure 12 shows such a gradient chromatogram. As demonstrated earlier, this chromatogram resolves a wide variety of impurities and degradation products for this product. One of the typical impurities is used as a system suitability sample to establish that the system is performing adequately for the determination in question. This application uses the isomeric derivative of cefaclor in which the double bond in the cephem ring is shifted one position. This entity is both a major degradation product of cefaclor and a potential synthetic impurity in the product. This compound does not have antimicrobial activity. The impurities are quantitated against cefaclor and specified impurity standards.

Gradient HPLC works well for high-tech laboratories found in major pharmaceutical companies and in the regulatory agencies of more sophisticated countries. Unfortunately, the pharmacopeial associations in many countries will not readily accept gradient HPLC procedures. Thus, alternative approaches must be developed.

Alternate approaches often use multiple isocratic systems to detect impurities and degradation. A weak isocratic system is used for the detection of the more polar components of the sample. Such a system is shown in Figure 13. A second isocratic

- Ease of use
- Ruggedness of procedure
- Transferability
- Ease of automation
- Regulatory requirements
- Pharmacopeial monograph
  requirements

Figure 10. Factors to be considered when developing impurity analysis methods.

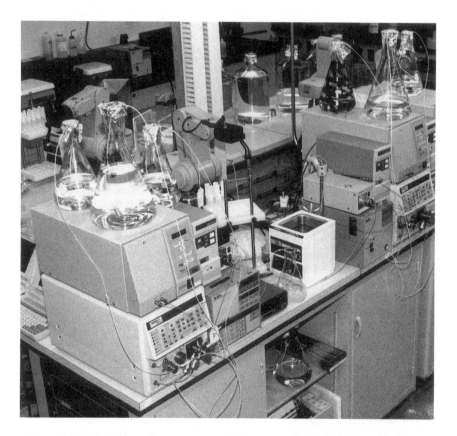

Figure 11. Robotic sample processor used to process dry samples for subsequent HPLC analysis.

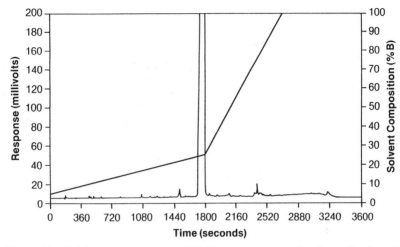

Figure 12. Gradient reversed phase HPLC chromatograms for determination of impurities. (See Figure 4 for conditions.)

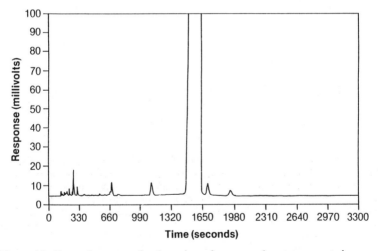

Figure 13. Isocratic system for detection of strong polar components in cefaclor. Mobile phase: 60 mL acetonitrile and 940 mL of a solution containing 1 gram of pentane sulfonic acid sodium salt per L, and 6.9 g of monobasic sodium phosphate adjusted to pH of 4.5 with sodium hydroxide. Column: Beckman Ultrasphere ODS (25 cm). Flow rate: 1 mL/min. Detection: Ultraviolet at 220 nm.

system using a stronger solvent system allows for the elution of less polar impurities and degradents that may be present in the sample. Figure 14 shows such a chromatogram. The evaluation of these systems shows very good agreement between the isocratic and gradient approach. The major difference between the two approaches is the inability of the isocratic system to detect the very small late eluters. Nevertheless, pharmacopeial associations accept the isocratic approach to screen a drug for impurities and degradents. The negative aspect of executing this type of dual assay procedure is the added time requirements for the total assay.

HPLC has become a major tool in the development of antibiotic products. HPLC methods are used to control products that are being marketed. Much of antibiotic product development uses high-tech procedures. Today, one must address the question of the best way scientifically to accomplish analytical goals in an manner compatible with capabilities present in routine control situations. Often alternative procedures must be developed to support pharmacopeial and regulatory requests when compounds are marketed in an international marketplace. One must also begin to consider the laboratory and control facilities that will be in place in the future.

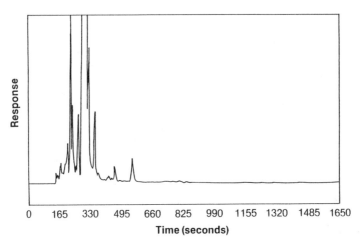

Figure 14. Isocratic system for detection of less polar components in cefaclor. Mobile phase: 150 mL acetonitrile and 850 mL of a solution containing 1 gram of pentane sulfonic acid sodium salt per L, and 6.9 g of monobasic sodium phosphate adjusted to pH of 4.5 with sodium hydroxide. Column: Beckman Ultrasphere ODS (25 cm). Flow rate: 1 mL/min. Detection: Ultraviolet at 220 nm.

Acknowledgments

The author wishes to recognize Ms. F. N. Bashore and Dr. S. W. Baertschi, who have made major contributions to the overall development program for cefaclor.

RECEIVED September 2, 1992

Chapter 6

Gas Chromatography and Pharmaceutical Analyses

Harold M. McNair and Ketan M. Trivedi

Department of Chemistry, Virginia Polytechnic Institute and State
University, Blacksburg, VA 24061

Gas chromatography (GC) is a separation technique involving an equilibrium of a sample in a vapor phase carried by inert gas through a column containing a stationary phase. The stationary phase consists of either a finely divided solid particles (adsorbent) or a high boiling liquid coated on solid particles (packed GLC column) or simply a thin film of liquid coated on the walls of the column tubing (capillary GLC Column). Since the sample has to be in the vapor phase to be transported, GC is limited to thermally stable volatile compounds. Pharmaceutical products are usually polar, water soluble organic compounds (water, or sea water is the basis for life on this planet). Water solubility of organic compounds is produced by the presence of one or more polar functional groups like carboxylic acids, amines, phenols and alcohols. In general it is these functional groups which have proven difficult to be handled by GC. Hence at first glance pharmaceutical analyses by GC may not seem applicable. Analyses of pharmaceuticals is however the single largest application area for gas chromatography. Table I shows the results of a 1989 survey of 1000 GC users conducted by LC/GC magazine.(1)

Table I - GC Application Areas

Field	Respondent (%)
Pharmaceuticals	16.7
Environmental	16.1
Organic Chemistry	14.6
Agriculture, food	13.5
Medical, biological	13.0
Plastics, polymers	7.8
Energy, petroleum	5.7
Forensics, narcotics	2.6

This widespread application of GC in pharmaceutical analyses can be explained by at least three major advantages of GC systems today. First, GC is a simple, rapid, quantitative and easily automated technique. Secondly, fused silica capillary GC columns, available today are so inert that highly polar molecules like free carboxylic acids and amines can be analyzed by current GC technology. And

0097–6156/92/0512–0067$06.00/0

finally, in the pharmaceutical industry, analyses by GC is not restricted to the final products. GC is widely used for the analyses of raw materials, solvents, intermediate products and organic pollutants in air, water, and finished products. In this chapter we will define GC as an analytical technique and give a general overview of recent advances that has attributed to its popularity in the pharmaceutical industry.

INSTRUMENTS FOR GAS CHROMATOGRAPHY

A wide variety of instruments differing in sophistication and price range (from $3,000 to $30,000) for GC are available on the market. The basic components of these instruments are schematically represented in Figure 1. They include : (1) gas cylinder containing inert carrier gas; (2) two-stage regulator; (3) flow control; (4) sample inlet; (5) column oven; (6) column; (7) detector; and (8) recorder or data handling system.

Chemically inert carrier gases (primarily helium, nitrogen and occasionally hydrogen) are used to transport the vaporized sample through the column. Typically helium is the most widely used. It will be shown later that the choice of gases is usually dictated by the type of detector employed. Associated with the gas supply are two stage pressure regulators, pressure gauges and flow controllers. In order to insure reproducible retention times (time required after sample injection for solute peak maximum to appear) and to minimize detector drift and noise it is necessary to carefully monitor and control the flow rate of carrier gas. This function is carried out by constant mass flow controllers for packed columns, constant pressure controllers for capillary columns, and most recently electronic pressure control to maintain constant flow rate in capillary columns under temperature programmed conditions. The sample is injected (usually with a micro syringe) into the heated injection port where it is rapidly vaporized and carried onto the column. Many pharmaceutical compounds are thermolabile and could decompose or rearrange under typical "hot injector" techniques. More recently a "cold on column injection" technique has been developed which places thermolabile compounds directly onto a cool column. Temperature programming is then used to move the sample through the column. Less thermal decomposition is observed for some samples.

Two types of columns are encountered in GC, packed and open tubular or capillary columns. Packed columns are usually 1/4 or 1/8 inch outside diameter stainless steel or glass tubes (2 to 3 m length) which are tightly packed with small solid support particles. A thin film (usually 3 to 10% by weight) of a high boiling liquid (stationary phase) is coated uniformly over the solid support. The solid support holds the liquid stationary phase in place so that a large surface area is exposed to the mobile phase; this results in a more efficient separation. Ideally the support should be inert small, uniform, preferable spherical particles with good mechanical strength. The sample partitions between the inert carrier gas and the liquid stationary phase into individual bands (peaks) based on differential solubility. Solutes with higher solubility spend more time in the liquid phase and move more slowly through the column. A properly prepared packed column would typically have 500 to 5000 theoretical plates (each equilibration of the solute between the mobile and stationary phase is defined as one thoretical plate) and may be used for several hundreds of analyses.

The open tubular or capillary columns were introduced by Golay in 1958. An open tubular column is not packed, it is simply a long thin tube (typical I.D. of 0.25 mm) with a very thin film (~0.2μm) of the stationary phase coated uniformly on the inside wall. From both theoretical and practical considerations it quickly became apparent that such columns could provide separations that were unprecedented both in resolution and speed. Early capillary columns were made of stainless steel and glass. Capillary columns since 1979 are made from fused silica tubing. Fused silica capillaries are commercially available with inside diameters of

100, 250, 320, and 530 μm. Their length ranges from 10 to 50 m and the corresponding number of theoretical plates range from 60,000 to 300,000 respectively. Thin films of stationary phase (0.2 to 1.0 um) are coated uniformly on the inner wall of the capillary column. These liquid phases are basically the same as those used for packed columns.

After the column, the carrier gas and the sample pass through the detector. The detector generates an electrical signal which passes to the recorder and generates a chromatogram (the written record of the analyses). In modern instruments a data handling system (integrator of P.C. based work station) automatically integrates the peak area, measures the retention time, performs calculations, and prints out a final report. A wide variety of detectors have been developed for GC analyses, however, only four have found widespread use. They are thermal conductivity (TCD), flame ionization (FID), nitrogen phosphorus (NPD), and electron capture (ECD). The application of these detectors in typical pharmaceutical analysis will be discussed shortly.

ADVANTAGES AND LIMITATIONS OF GC

One survey estimates that the total world sales of GC was $800 million in 1990. Table II highlights the major advantages of GC responsible for this popularity.

TABLE II - MAJOR ADVANTAGES OF GC

1. HIGH RESOLUTION - 100,000 theoretical plates
2. SPEED - many analyses complete in minutes
3. SENSITIVITY - ppm capability for almost all
    samples, ppb, capability with selective detectors
4. ACCURATE QUANTITATIVE RESULTS
5. WELL KNOWN, STRAIGHT FORWARD
    TECHNIQUE

The first advantage of GC is high resolution. One obvious way of increasing resolution is to increase the number of theoretical plates. Using 50m capillary columns one can easily generate 100,000 to 300,000 theoretical plates. A theoretical plate is a measure of goodness of the column; it is defined as one equilibrium between the mobile and stationary phases in the column. This large number of plates results in better resolution of complex mixtures by GC than by any other separation technique. One good example is the resolution of over 160 peaks from the aroma (head space) of coffee by capillary GC.

The second major advantage is the speed. Most separations (represented as chromatograms) are complete in matter of minutes. However, analyses of simple mixtures by GC is possible in seconds. For example Figure 2 shows the chromatogram of benzene, toluene, and ortho-xylene made on a short capillary column (6m in length), coupled with a fast flow of hydrogen carrier (180cm/sec). Note that high speed GC (seconds) is still at exploratory stages, hence, for the scope of this chapter which focuses on pharmaceutical analyses we shall limit our discussion to conventional GC techniques.

Sensitivity is the third advantage. With FID, concentrations in ppm (even sub ppm) range is routinely reported for volatile organic compounds. If selective detectors like ECD or nitrogen phosphoros (NPD) are used concentration levels as

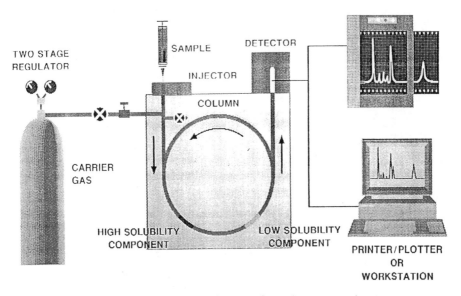

Figure 1. Schematic diagram of gas chromatograph.

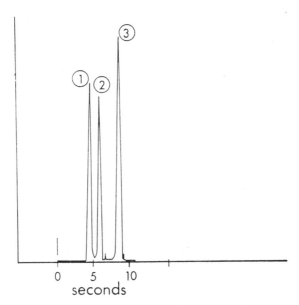

Figure 2. Chromatogram of (1) benzene, (2) toluene, (3) o-xylene by high speed GC analyses. Capillary column length 6m, carrier gas H_2, 180 cm/sec.

low as ppb (part per billion) have been reported. Trace amounts of contaminants in air, water, solvents or physiological fluids are routinely analyzed by GC because of the wide variety of both selective and sensitive detectors available.

Accurate quantitative analyses is another important feature of GC. GC offers the possibility to perform accurate quantitative analyses over a wide range of concentrations from milligram to picogram levels. Automated and reproducible computer control of sample introduction, carrier gas flow rate, column temperature and integrator analyses of the data has resulted in levels of accuracy and precision not thought possible ten years ago. Typical RSDs of 1% have been reported down to 0.1 ppm (100 ppb) for pesticide formations.

Finally GC is a well established, straight forward technique. Most BA/BS chemists work with a GC system in undergraduate school. The modern GC's are easy to operate, and the training required is reasonable compared to other techniques like high performance liquid chromatography (HPLC) or gas chromatography-mass spectroscopy (GC/MS), or headspace or purge and trap GC.

GC as a technique suffers from one important drawback. GC techniques, by themselves, cannot confirm the identiy of any given peak. For qualitative analysis by GC one typically uses retention times or Kovats Retention Indices.*(2)* Each component has a characteristic and reproducible retention time under carefully controlled conditions.*(3)* The problem is that these retention times are not unique, several components can have very similar retention times. An improved chromatographic method is the use of Kovats Retention Indices. These are retention times referenced to the elution of n-paraffin standards before and after the peak of interest. These relative retention times can be accurately measured to three significant figures and are usually independent of small changes in flow rate or column temperature. For confirmation of peaks spectroscopic techniques like mass spectroscopy (GC/MS) or fourier transform infrared spectroscopy (FTIR) are coupled to the GC. These hyphenated techniques are expensive and often require skilled operators, but they aid considerably in the identification of unknown peaks.

ADVANCES IN CAPILLARY GC RELATED TO PHARMACEUTICALS

Capillary column requires small sample sizes (~10^{-5}g or less) and low carrier gas flow rates (~ 1ml/min). These pose problems for sample introduction techniques. There are three main types of systems available commercially to inject samples onto capillary columns: split, splitless and on-column. Each have their advantages and limitations. The first two are well described in a recent text. The most gentle technique for sample introduction is a cold on column injection. A regular syringe is used to inject small quantities onto a cool column, and later temperature programming is used to elute the peak. This technique is gaining more popularity particularly in drug analyses even though it is an expensive accessory (~\$4K).

The use of capillary GC for pharmaceutical analyses and drug related research has grown rapidly in the last decade. Numerous drugs and their metabolites can be analyzed by GC after conversion to suitable volatile derivatives. A few routine and well established derivatives include silyl, methyl, and heptafluorobutyryl derivatives. Highly polar carboxylic acid, phenolic, and amine functionality are converted to more volatile esters, ethers and substituted amines.

Exotic separations, for example chiral separations of individual optical isomers using optically active silicone polymers coated on glass capillary columns have also been reported.*(4)* Figure 3 shows the example of a capillary GC technique resolving two optically active forms of norephedrine.

The underivatized forms of estrone (ES), equilin (EQ), equilenin (EQN), 17 alpha-estradiol (ESD), 17 alpha-dihydroequilin (DHEQ) and 17 alpha-dihydroequilenin (DHEQN) have been analyzed by capillary GC.*(5)* Determinations of these raw pharmaceuticals was performed following acidic hydrolysis of their sodium sulfate esters. Figure 4 depicts a representative chromatogram of a standard mixture containing equal amounts of mestranol(IS) and the equine estrogen. The method has the advantages of the GC technique itself which include simplicity and reproducibility and trace analysis capability.

Koves and Wells have described a GC-nitrogen phosphorous detector suitable for qualitative and quantitative analyses of basic drugs in postmortem blood samples.*(6)* A splitless mode of injection was employed with emphasis on assessing capillary GC in terms of sensitivity, resolution and reproducibility for qualitative and quantitative analyses of blood extracts. Figure 5 outlines the extraction procedure developed to eliminate the endogenous blood impurities. This 'sample preparation' step is essential for "dirty" matrices like blood, urine and body fluids. Two capillary columns differing in polarity, namely, DB-1701 and DB-1 were used in this analyses. Figure 6 shows the chromatograms of blood extract using these columns. A distinct advantage of using two columns for qualitataive analysis is observed. Doxepin and pyrilamine elute at the same retention time on the DB-1 column, while on DB-1701 these compounds are well resolved.

Figure 7 shows a drug screening procedure used for monitoring "illegal" drugs commonly found in athletes. In the 1980 Winter Olympic Games at Lake Placid GC was the official testing procedure.*(4)* It has been adopted and used in all subsequent Olympic games. Preliminary screening is made by GC, and confirmation of positives by GC/MS.

DRUG SCREENING USING TWO COLUMNS

If one uses two different liquid phases, the reliability of GC retention times as a means of qualitataive analysis improves greatly. It is a widely accepted practice in drug screening labs that SE-30 and OV-17 are the preferred liquid phases. Both are high temperature stable silicone polymers, but they differ in polarity sufficiently to be good complements.*(7)*

Table III gives the data concerning the retention indices and associated standard deviations for twenty compounds for which more than ten values using SE-30 were available. Comparable data for retention indices measured on OV-17 have also been included for comparison.

The chromatographic peak area (or height) is proportional to the quantity of the compound eluting from the column. To determine the area under a chromatographic peak one commonly uses an electronic integrator.

Quantitating residual solvents in bulk pharmaceuticals is a serious problem owing to the increasingly strict regulations on contaminants.*(8)* GC analyses of pharmaceuticals and their corresponding metabolites in blood etc. require a carefully designed sample preparation step. A variety of chromatographic techniques ranging from dynamic*(9)* or static*(10)* headspace GC to the use of purge and trap systems have been developed for analyzing volatile solvents. Most of these techniques are used as generalized methods for the determination of a wide variety of residual solvents in pharmaceutical compounds. Although most examples described here have minimal matrix interference, however, interferences in a specific determination by a drug or its thermal decomposition products are always a possibility. Normally

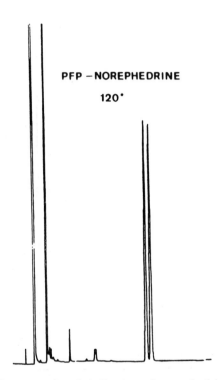

Figure 3. Pentafluoropropionyl derivative of nozephedrine separated as enantiomeric pair on an 18m x 0.25mm ID glass capillary column cooled with an optically active staionary phase.

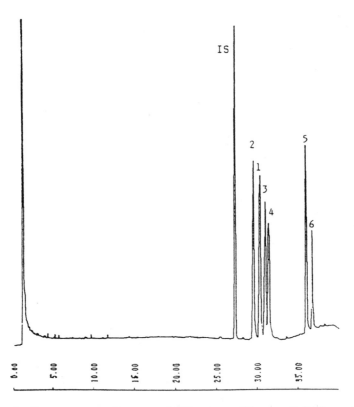

Figure 4. Chromatogram of mestranol (IS), estrone (1), estradiol (2), equilin (3), dihydroequilin (4), equilevin (5) and dihydroequilenin (6) standard mixture separated on a 25m x 0.32mm ID CP-Sil-5-CB column coupled with a 10m x 0.32mm ID CP-Sil-19-CB column.

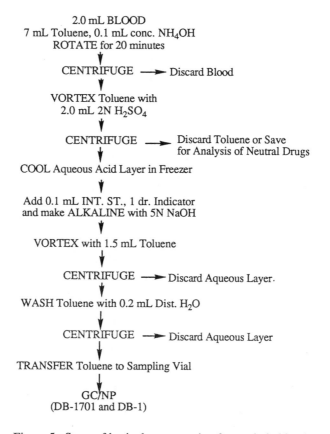

Figure 5. Steps of basic drug extraction from whole blood.

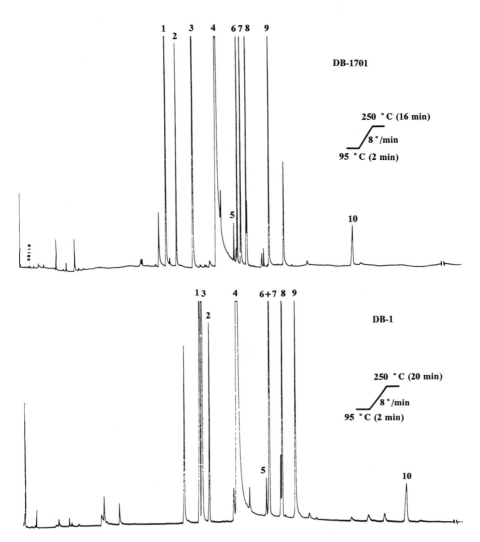

Figure 6. Chromatogram of a case blood extract - (1) pheniramine, (2) diphenhydramine, (3) caffeine, (4) metoprolol, (5) Cis-dosepin, (6) trans-doxepin, (7) pyrilamine, (8) SKF-525A, (9) codeine and (10) flurazepam.

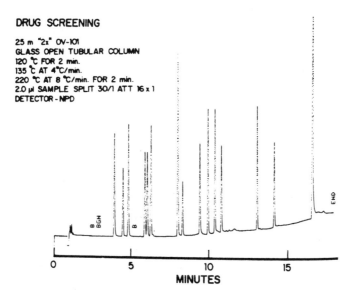

DRUG SCREENING

25 m "2x" OV-101
GLASS OPEN TUBULAR COLUMN
120 °C FOR 2 min.
135 °C AT 4°C/min.
220 °C AT 8 °C/min. FOR 2 min.
2.0 µl SAMPLE SPLIT 30/1 ATT 16 x 1
DETECTOR - NPD

Time'	Area	BC	RRT	RF	C	Name
4.04	3.0654	V	0.301	0.856	25.0093	Amphetamine:
4.62	1.9748	V	0.345	1.328	24.9952	Phentermine:
4.97	3.2924		0.371	0.797	25.0102	Methamphetamine:
6.06	2.3726	V	0.452	1.106	25.0105	Fentluramine:
6.21	2.4752	V	0.464	1.060	25.0072	Fludorex:
6.45	3.2755	V	0.482	0.801	25.0062	Mephemtermine:
8.22	5.3203	V	0.614	0.493	24.9988	Nicotine:
8.53	1.4601	V	0.637	1.796	24.9944	Chlorphentermine:
9.67	2.7955	V	0.722	0.938	24.9920	Methylephedrine:
10.20	3.2433	V	0.762	0.809	25.0080	Phenmetrazine:

Figure 7. Capillary GC separation on a 25m "2x" OV-101 column of various drugs and their detection with a nitrogen-phosphorous detector.

TABLE III: Retention indices (RI), standard Deviations (SD) and retention index references (ΔRI) for twenty compounds for which multiple determinations on both SE-30 and OV-17 stationary phases were available

Compounds	SE-30		OV-17		ΔRI
	RI	SD	RI	SD	
Amitriptyline	2196	22.13	2518	23.39	322
Amylobarbitone	1718	15.33	1988	10.97	270
Barbitone	1497	14.25	1796	13.01	299
Caffeine	1810	13.22	2246	23.82	436
Chlorpromazine	2486	31.56	2890	18.03	404
Cyclobarbitone	1963	12.30	2351	5.73	388
Dextropropoxyphene	2188	22.36	2455	25.53	267
Blutethimide	8136	17.62	2205	24.10	369
Lignocaine	1870	15.48	2164	26.19	294
Meprobamate	1796	14.51	2182	12.43	386
Methapyrilene	1981	11.70	2296	22.41	315
Methaqualone	2125	15.18	2580	19.85	455
Methylpheno-barbitone	1891	16.05	2262	25.73	371
Nicotine	1348	18.65	1553	15.60	205
Pentazocine	2275	22.20	2607	19.35	332
Pentobarbitone	1740	13.46	2017	14.35	277
Pethidine	1751	12.90	1996	9.92	245
Phenacetin	1675	16.25	2040	8.02	365
Phenobaarbitone	1957	13.22	2372	16.98	415
Salicylamide	1455	16.46	1802	16.19	347

interferences are suspected by the presence of extraneous chromatographic signals or some other anomalous behavior.

Drugs of abuse such as barbiturates, opiates, and depressants are one of the most common analyses performed by gas chromatography (pesticide residue is another).

Barbiturates. The widespread misuse of barbiturates has made it necessary for forensic laboratories to identify all common barbiturates. Because barbiturates have high polarity and low volatility, a GC column must possess thermal stability, high efficiency, good inertness, and selectivity. Barbiturates are typically chromatographed on a non-polar column (Rt_x-1) for primary screening and an intermediate polarity column (Rt_x-50) for confirmation.

Figure 8 shows the analysis of eleven common barbiturates on a 30m, 0.25mm ID, 0.25μm Rt_x-1, and Rt_x-50 run under identical conditions. The Rt_x-1, (dimethal silicone) used as the primary column, provides baseline resolution of the barbiturates in twelve minutes. The Rt_x-50, used as confirmation to the Rt_x-1, both provide baseline resolution in under twenty-three minutes. A dual column systems offer immediate confirmation on the two liquid phases of different polarity.

Opiates. The opiate drug class primarily refers to morphine and codeine, but is loosely used to describe a group of narcotic analgesics and their semi-synthetic derivativds. Opiate quantitation is typically performed on a non-polar column (Rt_x-1) or an intermediately polar column (Rt_x-50).

Figure 9 shows the simultaneous analysis of five opiates on a 30m, 0.25mm ID, 0.25μm Rt_x-1 and Rt_x-35. The Rt_x-1 and Rt_x-35 give baseline resolution between the critical components codeine, ethylmorphine, and morphine in under twenty-four minutes and show excellent peak symmetry.

Depressants. Depressants are a group of structurally similar compounds used as sedatives and abused for their euphoric properties. They are all weak bases and are typically analyzed on an intermediately polar column by FID.

Figure 10 shows the simultaneous analysis of five depressants on a 30m, 0.25mm ID, 0.25μm Rt_x-1 and Rt_x-35 at 260°C isothermal. Both columns give baseline resolution in 12.5 minutes and 30.5 minutes respectively. Because the depressants are highly polar, they are retained longer on the Rt_x-35 than on the Rt_x-1.

GC/MS

The tremendous analytical potential of the combined gas chromatograph/mass spectrometer (GC/MS) was first realized in 1957. The combination of these two techniques, abbreviated GC/MS has resulted in greater utilization of both because of the excellent manner in which they compliment each other. Both utilize small samples (10^{-6}g) and have good limits of detection (10^{-10}g). Where they differ is that GC has good separating ability but is a poor qualitative tool, while simple MS is difficult to use with complex mixtures but is excellent for confirmation of peak identity.

Typical detection techniques in MS include total ion current measurement and selected ion monitoring. Total ion current is basically a measure of the total number of ions formed in the ionization chamber from material eluting from GC. Figure 11 provides a comparison between a gas chromatogram detected by FID and a total ion current (TIC) detected by the mass spectrometer.*(11)* Notice there is little difference in the chromatogram. In selected ion monitoring of the molecular ion

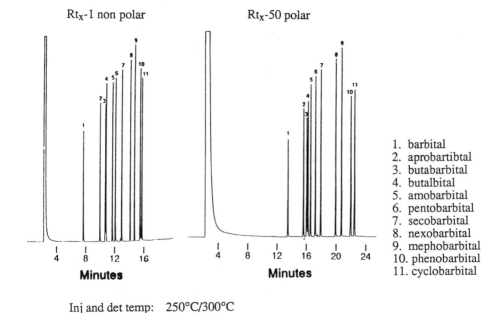

Rt$_x$-1 non polar   Rt$_x$-50 polar

1. barbital
2. aprobartibtal
3. butabarbital
4. butalbital
5. amobarbital
6. pentobarbital
7. secobarbital
8. nexobarbital
9. mephobarbital
10. phenobarbital
11. cyclobarbital

Inj and det temp:	250°C/300°C
Carrier gas:	He, 20 cm/sec
Oven Temp:	150°C for 5 min, 8°C to 290°C, hold for 5 min
Sample:	2μl, split 10/1
Column:	30m, 0.25mm ID, 0.25μm

Figure 8. Common barbiturates on 50m capillary column.
(Chromatogram courtesy of the Restek Corporation.)

Rt-1 non polar   Rt$_x$-35 polar

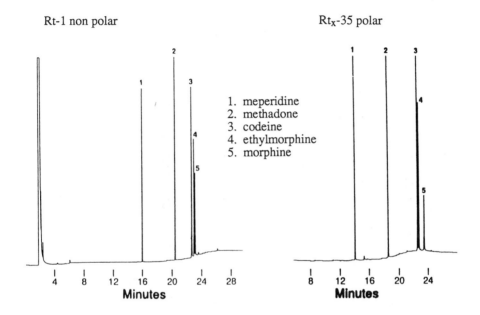

1. meperidine
2. methadone
3. codeine
4. ethylmorphine
5. morphine

Inj and det temp:	250°C/300°C
Carrier gas:	He, 20 cm/sec
Oven temp:	100°C at 10°C/min to 3000°C, hold 15 min
Sample:	2µl, split 10/1
Column:	30m, 0.25mm, 0.25µm

Figure 9. Confirmational analysis of opiates on nonpolar and polar liquid phases. (Chromatogram courtesy of the Restek Corporation.)

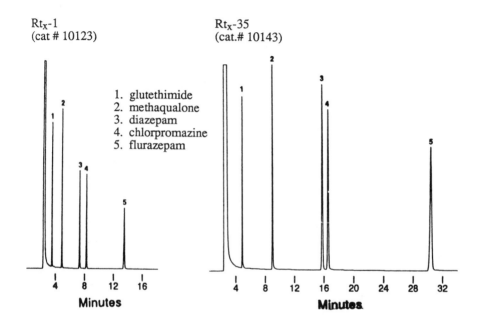

Rtx-1
(cat # 10123)

Rtx-35
(cat.# 10143)

1. glutethimide
2. methaqualone
3. diazepam
4. chlorpromazine
5. flurazepam

Minutes

Minutes

Inj and det temp:	260°C/300°C
Carrier gas:	He, 20 cm/sec
Oven Temp:	260°C isothermal
Sample:	2μl, split 10/1
Column:	30m, 0.25mm ID, 0.25μm

Figure 10. Confirmational analysis of depressants on two dissimilar liquid phases. (Chromatogram courtesy of the Restek Corporation.)

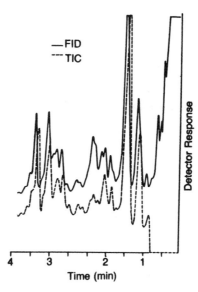

Figure 11. Simultaneous detection by flame ionization detector (FID) and total ion current (TIC) in a mass spectrometer of a urine sample.

(M^+) of carbamazepine with the molecular ion of dihydrocarbamazepine.*(12)* This is an example of a multiple ion monitoring where the intensities of two or more ions are recorded as a function of time.*(13)*   In this case, the dihydro compound is added as an internal standard for assaying blood levels of carbamazapine. Although the two compounds are not completely resolved by GC, their molecular ions can be resolved by MS. And finally, a computer system provides automated acquisition, normalization and plotting of the spectra. Typically one ends up with several hundred spectra stored in the computer, where they could be recalled to aid in the identification of peaks. Hence, in computer-reconstructed chromatograms one does not have to wait until a peak elutes to scan the mass spectrometer. Figure 12 shows a total ion chromatogram as a reconstructed spectrum resulting from the deconvolution of components of a urine extract.*(14)* By and large, such resolved spectra provide very good matches when library searches are used for their identification.

The mass spectrometer provides a large amount of qualitative information on very small samples. Since any compound that passes through a GC will be converted into ions in the MS, it is safe to say that MS is a universal detector for GC. Suitability for quantitative measurements has been reported, and sensitivity is competitive with ECD as shown in Table IV.

Table IV. Sensitivity of GC Detectors

Total ion current monitoring (MS)	10^{-8}g
Mass chromatogram	10^{-8}g
Flame ionization	10^{-10}g
Electron capture	10^{-12}g
Selected ion monitoring	10^{-12}g

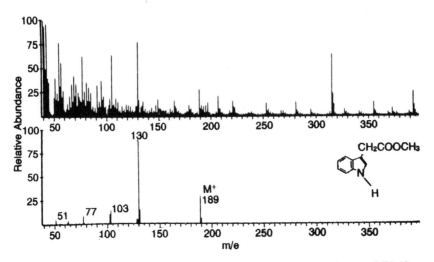

Figure 12. Mass spectrum of indole acetic acid 3-methylester from a GC/MS analysis of human urine (top) and the same spectrum resolved by automated data processing (bottom).

Literature Cited

1. LC/GC Magazine 1991, 8 #6, p. 442.
2. Kovats, E. Helv. Chim Acta 1958, 41, p. 1915.
3. Sprouse, J. F.; Varano, A. Amer. Lab 1989, 16(9), p. 54.
4. Novotny, M. Drug Metabolism Reviews 1981, 12(2), pp. 279-292.
5. Novakovic, J.; Tuzicka, E. J of High Resolution Chrom. 1991, 14, p. 495.
6. Koves, E.M.; Wells, J. Journal of Forensic Sciences 1985, 30, 3, pp. 692-707.
7. Ardrey, R.E.; Moffat, A. Journal of Chromatography 1981, 220, pp. 195-252.
8. Haky, J.E.; Stickney, T.M. Journal of Chromatography 1985, 321, pp. 137-144.
9. Wampler, T.P.; et. al. Pittsburg Conference on Analytical Chemistry, Atlantic City, New Jersey 1984, paper no. 604.
10. Bicchi, C.; Bertoline, A. Farmaco Ed. Pr. 1982, 37, p. 88.
11. Holmes, J.C.; Morrel, F.A. Appl. Spectrosc. 1957, 11, p. 86.
12. Fenselau, C. Anal. Chem. 1977, 49(6), p. 563A.
13. Sweeley, C.C.; et. al. Anal. Chem. 1966, 38, p. 1549.
14. Dromey, R.S.; et. al. Anal. Chem. 1976, 48, p. 1368.

RECEIVED May 1, 1992

Chapter 7

Gas Chromatography and Mass Spectrometry of *Erythrina* Alkaloids from the Foliage of Genetic Clones of Three *Erythrina* Species

Lori D. Payne[1] and Joe P. Foley[2]

Department of Chemistry, Louisiana State University,
Baton Rouge, LA 70803

The foliage of genetic clones of the nitrogen-fixing trees *Erythrina berteroana*, *E. poeppigiana* and *E. costarricensis* was evaluated by gas chromatography/ mass spectrometry for toxic alkaloid content. The major alkaloid identified in all three species was ß-erythroidine, a naturally derived drug used in the 1940s and 1950s as a neuromuscular blocking agent in surgery and electroshock treatments. Other biologically active alkaloids identified were α-erythroidine, erybidine, erythraline and various minor aromatic *Erythrina* alkaloids. Although all clones tested contained some amount of ß-erythroidine, a few clones were notably lower in this drug. Several clones contained a variety of alkaloids not present in all clones or in all species. The results indicate that the expression of *Erythrina* alkaloids is at least partially under genetic control and the differences in alkaloid composition between clones of *E. berteroana* and *E. poeppigiana* are more quantitative in nature rather than qualitative.

Extracts from *Erythrina* species were prepared by South American natives in concentrated form for use in poison arrows (*1*). The resultant muscle paralysis is caused by blocked neural transmissions at the myoneural junction, however, acetylcholine esterase is not involved in the pharmacological action. The principle active component in these extracts was ß-erythroidine (Figure 1) which was isolated from the seeds of *Erythrina americana* by Dr. Folkers of Merck and Co. in 1937 (*2*). Folkers and Unna (*3, 4*) began surveying the genus and they found that at least half of the species were biologically active. In the decade that followed additional types *Erythrina* alkaloids were isolated from a variety of *Erythrina* species and structure determination studies ensued (*5-10*).

[1]Current address: Analytical Research, Merck Sharp and Dohme Research Laboratories, P.O. Box 2000 (Ry80L−123), Rahway, NJ 07065
[2]Current address: Department of Chemistry, Villanova University, Villanova, PA 19085−1699

Because of its muscle paralyzing activity, ß-erythroidine was used in surgery and electroshock treatments in the 1940s and 1950s. It was the preferred drug over previously available quaternary type alkaloids because it could be administered orally as well as intravenously (11).

Folkers and Major (2) determined the LD50 of erythroidine hydrochloride in white mice to be 120 mg/kg when given perorally and 15 mg/kg when given subcutaneously. Complete motor paralysis resulted when 0.1-0.15 mg per frog was injected into the lymph sac (intralymphatic injection).

The most frequent use of ß-erythroidine was in controlling seizures and convulsions in the medical treatment for schizophrenia and other mental diseases (12). Cottington (13) describes the use of ß-erythroidine in treatment of children with schizophrenia to reduce the severity and duration of the metrazol induced seizures and to alleviate the dread associated with the treatment. Previously the treatment almost always resulted in compression fractures of the thoracic vertebrae. The administration of 100 mg of ß-erythroidine intravenously resulted in complete muscle relaxation and subsequent administration of metrazol produced mild or no seizures. This treatment was given three times a week. ß-Erythroidine was used in the same capacity in electroshock treatments (14).

Curare and curare like compounds were used in surgery to produce a degree of anesthesia impossible to obtain with nitrous oxide or ethylene anesthesia (15). ß-Erythroidine was attractive for this purpose because it could be administered orally or intravenously whereas curare could only be administered intravenously. ß-Erythroidine also had a greater margin of safety than did curare (16), although ß-erythroidine's margin of safety would be considered small by today's standards; the LD50 was twice the paralyzing dose in mice (16, 17). In the 1940s and 1950s, curare was widely used in American operating rooms. The use of ß-erythroidine and investigation of its mode of action were inhibited at that time by its unknown structure and by the induction of hypotension and bradycardia in some patients. The structure of ß-erythroidine was not correctly elucidated until 1953 by Boekelheide (18); soon after that it was replaced by more effective synthetic drugs.

Erythrina is an important tree species distributed worldwide at low to mid elevations. It is important in Latin America, Africa, Australia and Asia. *Erythrina* is one of many leguminous trees which are used in agroforestry systems, especially in Central and South America, as living fences, shade for cash crops such as coffee or cocoa, windbreaks, support for climbing crops, and more. The widely distributed *Erythrina* genus has been used in agricultural systems since first described by taxonomists in Central America. Its high nitrogen content is exploited by farmers in coffee and cocoa plantations where the shade species is pruned and the foliage left on the ground to provide nitrogen to the cash crop.

An additional use of the nitrogen fixing capabilities of *Erythrina* has been proposed by researchers. The proposed new use *Erythrina* is as a diet supplement for ruminant animals just as legumes are used as supplements in the temperate zones (19). Through microbial fermentation, ruminants have the capability of utilizing various types of organic matter as food. Microorganisms in the rumen of bovines, for example, obtain energy and protein from forage unusable by non-ruminants. They in turn produce microbial protein, B vitamins and other valuable nutrients that can be utilized by the bovine (20). Frequently, the limitation in the growth and quality of the animals is the amount and the availability of digestible nitrogen present in the feed. Many farmers supplement the pasture diet of ruminants with a high energy/high protein concentrate. However, for the small subsistence farmer, the cost of purchasing concentrate is prohibitive. As an alternative for the small farmer, other materials high in nitrogen, such as *Erythrina*, are readily available and may serve just as well as a concentrate in supplying high quality protein to ruminant animals. *Erythrina* contains 24.0 to 38.4 % crude protein in the leaves, 70% of which is digestible by ruminants (21). When present in a coffee agroforestry system the *Erythrina* tree may produce up to 90 kg of dry leguminous foliage per year per hectare (*22*).

Feeding studies with *Erythrina* have shown that the use of *Erythrina* foliage as a diet supplement is a possible alternative to concentrate (*23*). Animals have been fed up to 100% of *Erythrina* foliage in their diet with no visible adverse affects. However, it is known that *Erythrina* foliage contains toxic alkaloids (*24*) and that these alkaloids may affect the animals or contaminate the meat and/or milk of the animals consuming *Erythrina* foliage. It is not known whether the expression of *Erythrina* alkaloids is under genetic or environmental control or whether there are varieties of *Erythrina* species with no or low concentrations of toxic alkaloids in the foliage.

Other species now in use as feeds for animals, such as alfalfa and canary grass, have been known to contain undesirable compounds (*25*). With time and careful breeding strategies, these compounds have been essentially eliminated from domestic species greatly improving the forage quality of these types of pastures species and forestalling the possibility of milk or meat contamination. At the Centro Agronómico Tropical de Investigación y Enseñanza (CATIE), Turrialba, Costa Rica, an arboretum of nitrogen fixing trees has been established (*26*). The Latin American Nitrogen Fixing Tree Arboretum contains plots of individual clones of *Erythrina* species collected throughout Central America. Thirteen of these clones were selected for this preliminary study of the content and composition of *Erythrina* alkaloids in the foliage of three different *Erythrina* species, *E. berteroana, E. poeppigiana* and *E. costarricensis*, to determine if the expression of these alkaloids varies between genetic clones and to identify low alkaloid content clones of *Erythrina* to be used in the development of feeding programs for small family farmers of Central America.

Experimental

Clonal Selection. *Erythrina* clones were selected on the basis of genetic quality, nutritional quality, and availability after consultation with scientists at CATIE. A summary of this data is presented in Table I.

Sample Collection and Preparation. Twelve individual trees representative of each clone in the Latin American Nitrogen Fixing Tree Orchard were separated into three replicate groups at random. Leaves were detached from the second or third branch of the tree, collected and weighed. Immediately upon arrival to the laboratory, the leaf blades were clipped from the petiole and weighed. Two 100 gram samples from each replicate were immediately frozen; one fresh frozen sample was extracted as fresh material and the other was lyophilized (Eyela Tokyo Tatakikai Freeze Dryer LFD1) for two to three days until dry. The remaining leaf blades were placed in aerated paper bags and dried at 60°C for 72 hours and then milled with a Thomas Wiley Model 4 Lab mill to a powder. All samples were collected in June or July of 1987 and transported to LSU after preparation.

Extraction. Typical alkaloid extraction procedures were used. Figure 2 illustrates the flow chart of extraction of a representative sample. The fresh material was extracted with 50% aqueous ethanol macerated with a Servall Omni Mixer (Sorvall Inc., Norwalk, Conn.). The lyophilized and dried samples were extracted using 95% ethanol in a Soxhlet device. Subsequent cleanup involved the liquid-liquid extraction of the alkaloids using changes in pH. The concentrated material from the Soxhlet extraction was dissolved in 2% H_2SO_4. This acidic mixture was cleaned up with a chloroform extraction. The acidic mixture was neutalized with $NaHCO_3$ and then brought to pH 10 with 25% NH_4OH. The basic solution was extracted with chloroform, dried over $NaSO_4$ and concentrated. The residue was dissolved in chloroform to approximately 1 mg residue to 1 mL chloroform for GC/MS analysis.

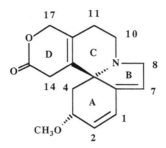

Figure 1. ß-Erythroidine.

Table I. Selected *Erythrina* Clones and Important Characteristics of Interest

CLONE # SPECIE[1]	PROVIDENCE (ORIGIN)	NUTRITIVE VALUE[2,3]	PHENOLOGY[2,4]	BIOMASS PRODUCTION[2,5]
E. berteroana[6]				
2674	Sarapiqui	med/low	superior	very high
2652	Naranjo	high	average	high
2670	Iroquois	high	superior	high
2667	La Suiza	high	superior	high
2689	Santa María	high	superior	med/low
2668	Paso Canoas	high	average	med/low
2691	Quebrada	low	superior	med/low
2703	Palo Verde	high	average	low
E. poeppigiana				
2661	Tejar	high	superior	med/high
2708	Alajuela	high	superior	med/high
2687	St.María Dota	high	superior	med/high
2700	San Pablo	high	average	medium
E. costarricensis[6]				
2750	Paso Canoas	high	average	n/a[7]

[1]Latin American Seed Bank Number, CATIE. [2]Perez, E.E., Master's Thesis in preparation, CATIE.. [3]Determined in consultation with María Kass, Animal Nutritionist, Animal Production Dept., CATIE. [4]Determined in consultation with Edgar Viquéz, Forestry Geneticist, NFTP, CATIE, and tree register information. [5]Nitrogen Fixing Tree Project (NFTP) evaluation. [6]Clone Numbers 2668, 2691 and 2703 were originally classified as *E. costarricensis*. [7]Not available

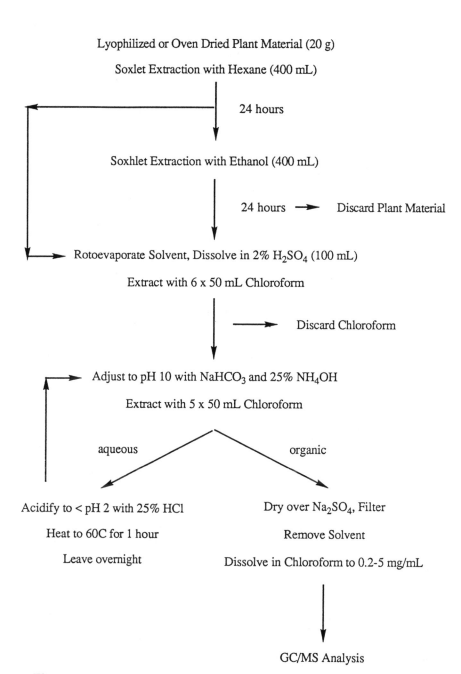

Figure 2. Extraction procedure for a representative plant material sample.

The aqueous portion left over from the final chloroform extraction above was hydrolysized by reacidifing to pH 2, heating for 1 hour at 60°C and leaving the solution overnight. This solution then was cleaned up, basified and extracted with chloroform as described above to give the hydrolysized alkaloid fraction.

Total Nitrogen Determination. Total nitrogen was determined using the microKjedahl method.

Statistical Analysis. Statistical analysis was performed using Statview 512+ (Abacus Concepts) on a Macintosh computer.

GC/MS Analysis. One microliter of the extract dissolved in chloroform was run on a Hewlett Packard 5890 gas chromatograph and analyzed with an Hewlett Packard 5971 quadrupole mass spectrometer using splitless injection. Gas chromatographic conditions were as follows: 40°C for 3 min., ramp of 20°C/min. to 250°C and held at 250°C for 25 minutes. The injector temperature was 200°C and the detector temperature was 280°C. A DB-5 (J&W Scientific) 30 m capillary column was used (0.25 mm i.d., 0.25 μm film thickness). The flow rate of helium was 30 cm/sec. The scan rate was 0.5 scans/sec and the mass range scanned was 15-550 mass units. Electron impact at 70 eV was used. The data was acquired and analyzed using DOS based ChemStation (Hewlett Packard) software.

Identification of *Erythrina* Alkaloids. In the case of ß-erythroidine, oxo-erythroidine, erysodine, erysonine, erysovine, erysopine, erythraline and erythratine, authentic standards were available for comparison purposes. Identification of these alkaloids was accomplished by retention time and mass spectra matching to the authentic standards which were run under identical conditions. Identification of erybidine and the unknown of m/z=254 was tentatively determined by comparison of the mass spectra relative ion intensity values obtained with those values found in the literature.

Results and Discussion

Other researchers who have used GC/MS to investigate *Erythrina* alkaloids have utilized the trimethylsilyl (TMS) derivatization technique with OV-17 columns (27-29). We found, at least for the alkaloids examined in this study, that TMS derivatization was unnecessary. Alkaloids with or without hydroxy groups were well resolved on a DB-5 column under the chromatographic conditions described in the experimental section. Figure 3 demonstrates this point. Erythratine and TMS derivatized erythratine are present in the same total ion chromatogram (TIC). Both compounds are easily detectable, albeit overloaded, with approximately a one minute difference in retention times. Other *Erythrina* alkaloids with free hydroxyl groups such as erysodine and erysovine were also detected with adequate sensitivity without TMS derivatization.

We were also concerned with alkaloid artifacts that oven-drying the leaf samples might introduce. Bhukani et al. (30) reported that some of the oxo *Erythrina* alkaloids may have been produced in the drying process . For this reason, all the samples in our study were divided into two groups, one which was lyophilized and one which was oven-dried. However, there was no qualitative difference in the alkaloid profile between the oven dried or lyophilized samples (Figure 4). 8-Oxo-erythroidine (retention time = 29.8 minutes) was present in samples processed by both methods. There was a difference, however, in the extraction efficiency between the two methods of sample preparation. Extraction of the oven-dried material yielded slightly more alkaloid extract by weight per gram sample than the extraction of the lyophilized samples. This may have occurred because the oven-dried samples were milled whereas the lyophilized samples were crushed resulting in finer particles in the oven-dried case.

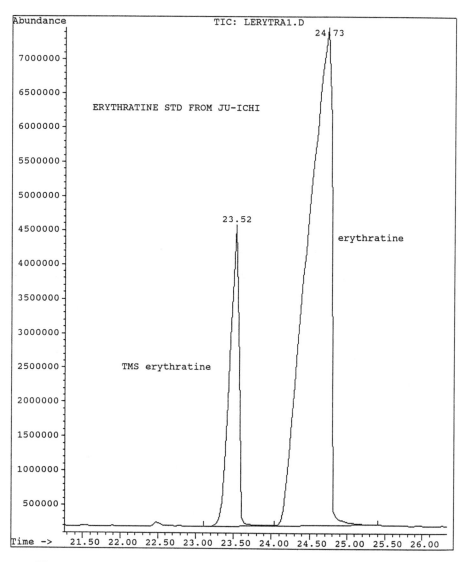

Figure 3. Total ion chromatogram (TIC) of erythratine and trimethylsilanized (TMS) erythratine.

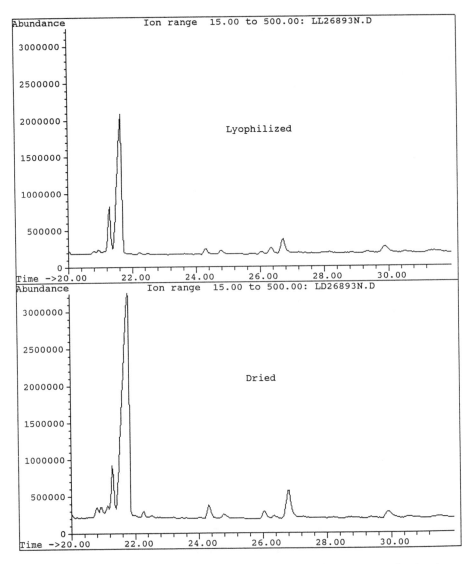

Figure 4. Total ion chromatogram (TIC) of extracts of plant material lyophilized and oven-dried at 60°C.

There has been quite a bit of controversy on whether the expression of secondary plant products such as alkaloids is under genetic or environmental control. It is commonly observed that the alkaloid composition of a plant may vary with environmental conditions such as soil nutrient level, soil type, soil pH, elevation, drainage, sunlight and stress, but no correlations have been detected using these parameters (*31-34*). On the other hand a correlation has been found between alkaloid content and the developmental stage of the plant (*35*).

In an effort to eliminate environmental variation in alkaloid composition in the present study, the clones utilized were grown under the same environmental conditions, although they originated from diverse regions of Costa Rica (Table I) and therefore are genetically distinct.

There have been attempts to use the presence of a particular alkaloid to taxonomically classify ambiguous plant specimens or species. In some genera, such as *Nicotiana*, the alkaloid composition was an unreliable parameter of phytogenetic position (*36*). On the other hand, Hargreaves et al. (*27*) were able to distinguish between old and new world *Erythrina* species by their alkaloid profiles. Our studies revealed that *E. poeppigiana*, *E. berteroana* and *E. costarricensis*, which are all members of the new world *Erythrina* group, are more similar than distinct with respect to the specific *Erythrina* alkaloids encountered and it may be difficult to classify these species by their alkaloid profiles.

All three of the above species tested in our laboratory belong to the American species of *Erythrina*. *E. poeppigiana* belongs to the most modern *Erythrina* taxonomic section, Micropteryx, while *E. berteroana* and *E. costarricensis* belong to the section *Erythrina*, which is still evolving (*35*). As a matter of fact, it is difficult to distinguish between *E. berteroana* and *E. costarricensis* and it is believed that these two species may interbreed (*8*). As a case in point, three clones originally classified as *E. costarricensis* (2668, 2691 and 2703) were reclassified as *E. berteroana* during the course of this study by the Nitrogen Fixing Tree Project at CATIE where the arboretum is located.

The alkaloids profiles of the thirteen clones are presented in Table II. All the clones investigated contained ß-erythroidine (ERTR) as the major alkaloid component. 8-Oxo-erythroidine (OXOE) was detected in half the *E. poeppigiana* and *E. berteroana* clones, but it was not present in *E. costarricensis*. Erysodine (ERSD) and was may be erybidine (ERBD) were detected in the majority of the clones. Erybidine has not been reported in these species previously and no authentic standard was available for verification. Tentative identification was based on comparison to literature mass spec values. Erybidine has been reported in *E. xbidwilli*, *E. crista galli*, *E. herbacea* and *E. senegalensis* which are Asian species of *Erythrina* (*37-39*). Erythraline (ERTL) and erythratine (ERTH) were detected in only a few cases.

Five compounds were detected in the clones that could not be identified. Several lower molecular weight compounds, two isomers with a molecular weight (MW) of 247, and one compound with a molecular weight of 245 eluted before the other *Erythrina* alkaloids with retention times between 15.5 and 16.5 minutes (Figure 5). The unknown of MW of 245 was almost ubiquitous in the samples analyzed. The 247 isomers were concentrated in *E. berteroana* primarily accompanied by unknown 245. The 247 isomers may be dihydro-derivatives of the 245 compound. From their mass spectra it is apparent that they contain a methoxy group ($(M-15)^+$ and $(M-31)^+$). Trimethylsilation revealed one hydroxy group in each of the unknowns. The fragmentation pattern and the presence of a peak at m/z of 130 suggests that these unknowns may be dienoid type *Erythrina* alkaloids as yet undescribed. Additional work is required in order to further elucidate the structures of these unknowns.

An unknown of molecular weight 254 matches the description of a new type of *Erythrina* alkaloid reported by Redha (*29*) which contains two nitrogens. The mass spectra are identical within experimental error, but an authentic standard of the alkaloid was unavailable for retention time matching via standard addition. The unknown of molecular weight 289 may be a dihydro-derivative of a known *Erythrina* alkaloid.

Table II. Alkaloid Profiles of Selected *Erythrina* Clones [1]

CLONE #	ERTR	OXOE	EYSD	ERTL	ERBD	ERTH	247A	247B	245	254	289
berteroana											
2652	*	*			*				*		*
2667	*	*	*				*		*		*
2670	*	*			*		*	*	*	*	*
2674	*	*	*	*	*	*	*	*	*	*	*
2689	*		*	*	*	*	*	*	*	*	*
26682	*		*	*	*		*	*	*	*	
26912	*		*	*	*	*	*	*	*	*	
27032	*		*	*	*		*	*	*	*	
poeppigiana											
2661	*	*	*		*				*		*
2687	*	*	*		*				*		*
2700	*		*	*			*		*	*	*
2708	*		*		*						
costarricensis											
2750	*				*				*	*	*

1ERTR=ß-erythroidine, OXOE = 8-oxo-ß-erythroidine, ERSO = erysodine, ERTL = erythraline, ERBD = erybidine, ERTH = erythratine, 247A, 247B, 245B, 254, 289 = unknowns.
2Originally classified as *E. costarricensis*, reclassified as *E. berteroana* July, 1991.

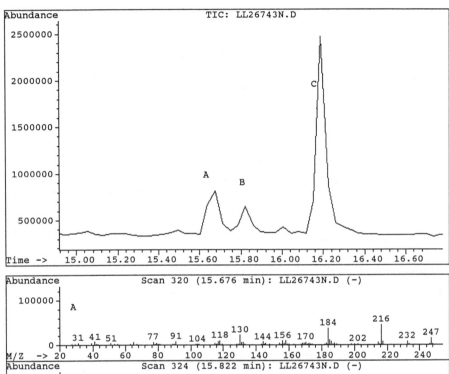

Figure 5. Total ion chromatogram (TIC) and mass spectra of three unknowns: (A) 247A, (B) 247B, and (C) 245.

Examination of the alkaloid profile reveals that the three reclassified clones (2668, 2691 and 2703) along with *E. berteroana* clone 2689 have alkaloidal characteristics similar to both *E. costarricensis* and the other *E. berteroana* clones. All four of these clones lack 8-oxo-erythroidine as does the one remaining *E. costarricensis* clone. However, they also contain the unknowns 247A and 247B which are not present in *E. costarricensis* but are common in *E. berteroana*. Alternatively, only 2689 and 2691 contain erythratine, a alkenoid type of *Erythrina* alkaloid and along with clone 2703, they contain erythraline, a dioxymethylene *Erythrina* alkaloid. It is possible that these four clones represent hybrids between *E. berteroana* and *E. costarricensis*. They were certainly the most interesting in terms of alkaloid composition of all the clones studied. To fully assess the utility of alkaloid profiles for taxonomic classification of *Erythrina* species, additional genetic clones have to be analyzed for alkaloid content, especially of clones *E. costarricensis* which is poorly represented in this study.

By far the alkaloid present in the greatest quantity in all three species was ß-erythroidine. ß-Erythroidine had been detected in the seed of *E. poeppigiana*, *E. berteroana* and *E. costarricensis*, but ß-erythroidine had been reported as a major alkaloid in the leaves of *E. poeppigiana* only (*40*). Romeo also reported that there was no correlation between amino acid content and alkaloid content in the species he tested. This was corroborated in our studies where there was no correlation ($r2=-0.014$) between total nitrogen content and ß-erythroidine content (Table III). The non-correlation of alkaloid content with nitrogen or protein content has been observed in other nitrogen fixing tree species such as *Leucaena* (*41*). This is fortunate from a genetic engineering standpoint because it suggests that *alkaloid content and nitrogen content are not genetically linked*. Moreover, *there is great variability between clones of the same species in ß-erythroidine content* (Table III). There are low alkaloid content clones of each species which might be preferred for use in the development of alternative animal feeds. This variability also suggests a means of breeding low alkaloid, high protein *Erythrina* varieties for animal feed. Importantly, variation between individuals of the same genetic clone was very small both in total nitrogen and ß-erythroidine content.

Once it was determined that ß-erythroidine was the major alkaloid in these three *Erythrina* species, it was not surprising to find very little variation in alkaloid composition *between the species*. ß-Erythroidine is formed biosynthetically by extra-diol type cleavage of the aromatic *Erythrina* alkaloids (*42*). The cleavage is probably extra-diol because of the occurrence of *Erythrina* alkaloids with a nitrogen at the C-16 position. An example of a dinitrogen *Erythrina* alkaloid is melanacanthine reported by Redha (*29*). Therefore, if ß-erythroidine is the final product of the *Erythrina* alkaloid biosynthetic pathway in the *Erythrina* species examined, then it follows that all possible biosynthetic precursors, namely the aromatic types of *Erythrina* alkaloids, are shunted into this pathway and are converted principally to ß-erythroidine, which is the major alkaloid detected. The aromatic *Erythrina* alkaloid precursors may not be present or detected in significant amounts because they may be rapidly converted to ß-erythroidine.

Conclusions

GC/MS using a DB-5 column is a valuable technique for the analysis of *Erythrina* alkaloids. Clonal evaluation of *E. poeppigiana*, *E. berteroana* and *E. costarricensis* revealed that ß-erythroidine, a drug previously used in surgery and electroshock treatments, was the major alkaloid in the leaves of these species. There was little statistical variability in the parameters measured between individual trees of the same genetic clone. There was wide variability in the concentration of ß-erythroidine between *Erythrina* clones of the same species and between different *Erythrina* species, however there was little difference in the types of *Erythrina* alkaloids detected between clones or between species. Low alkaloid content clones of all three species were identified.

Table III.   Alkaloid and Nitrogen Content of thirteen *Erythrina* Clones

CLONE #	SPECIES	ß-ERYTHROIDINE mg/100g fresh wt.	TOTAL NITROGEN percent dry wt.
VERY HIGH			
2652	berteroana	114.4	4.28
HIGH			
2667	berteroana	65.2	4.92
2670	berteroana	52.7	5.05
2708	poeppigiana	46.0	4.68
AVERAGE			
2703	berteroana	23.2	4.50
2661	poeppigiana	20.5	5.43
2689	berteroana	20.0	4.44
MODERATELY LOW			
2674	berteroana	11.2	4.81
2687	poeppigiana	10.0	5.86
2668	berteroana	9.2	4.39
2691	berteroana	7.1	4.39
LOW			
2700	poeppigiana	2.2	5.08
VERY LOW			
2750	costarricensis	0.5	4.48

Acknowledgments

The authors thank CATIE and the personnel of the Nitrogen Fixing Tree Project and the Animal Nutrition Laboratory for their assistance. CATIE also contributed to the cost of the collection of the *Erythrina* clones. Gifts of ß-erythroidine from Dr. V. Boekelheide of the University of Oregon, erysodine and erysovine from Dr. A. H. Jackson of the University of Cardiff, and erythraline and erythratine from Dr. M. Jui-ichi of the Mukogawa Women's University in Japan are gratefully acknowledged.

Literature Cited

1. Craig, L.E. *The Alkaloids: Pharmacology*; Academic Press: New York, NY, 1955; Vol. 5.
2. Folkers, K. Major, R.T. *J. Am. Chem. Soc.* **1937**, *59*, 1580-1581.
3. Folkers, K.; Unna, K. *J. Pharm. Sci.* **1938**, *27*, 693-699.
4. Folkers, K.; Unna, K. *J. Pharm. Sci.* **1939**, *28*, 1019-1028.
5. Folkers, K.; Koniuszy, F. *J. Amer. Chem. Soc.* **1940**, *62*, 436-441.
6. Folkers, K.; Koniuszy, F. *J. Amer. Chem. Soc.* **1940**, *62*, 1673-1676.
7. Folkers, K.; Koniuszy, F. *J. Amer. Chem. Soc.* **1940**, *62*, 1677-1683.
8. Folkers, K.; Shavel, J. jr.; Koniuszy, F. *J. Amer. Chem. Soc.* **1941**, *63*, 1544-1549.
9. Folkers, K.; Shavel jr., J. *J. Amer. Chem. Soc.* **1942**, *64*, 1892-1896.
10. Koniuszy, F.; Wiley, P.F.; Folkers, K. J. Amer. Chem. Soc. **1949**, *71*, 875-878.
11. Megirian, D.; Leary, D.E.; Slater, I.H. *J. Pharmacol. Exper. Therap.* **1955**, 113, 212-277.
12. Lehman, A.J. *Proc. Soc. Exptl. Therap.* **1936**, *33*, 500-503.
13. Cottington, F. *Am. J. Psychiatry* **1941**, *98*, 397-400.
14. Robbins, B.H.; Lundy, J.S. *Anesthesiology* **1947**, 8, 348-357.
15. Robbins, B.H.; Lundy, J.S. *Anesthesiology* **1947**, 8, 252-265.
16. Berger, F. M.; Schwartz, R.P. *J. Pharmacol. Exper. Therap.* **1948**, *93*, 362-367.
17. Sauvage. G.L.; Boekelheide, V. *J. Amer. Chem. Soc.* **1950**, *72*, 2062-2064.
18. Boekelheide, V.; Weinstock, J.; Grundon, M.F.; Sauvage, G.L.; Agnello, E.J. *J. Amer. Chem. Soc.* **1953**, *75*, 2550-2558.
19. Benavides, J.E. *Curso Corto Agroforestal*; CATIE: Turrialba, Costa Rica, 11-21 January, 1983.
20. Hungate, R.E. *The Rumen and Its Microbs*; Academic Press: New York, NY, 1966.
21. Espinoza Bran, J.E. *Characterización nutritiva de la fracción nitrogenada del forraje de madero negro Gliricidia sepium y poró Erythrina poeppigiana*, Master's Thesis; CATIE: Turrialba, Costa Rica, 1984.
22. Russo, R.O. *Erythrina: Un Genero Versatil en Sistemas Agroforestales del Tropico Húmedo*; INFORAT-CATIE: Turrialba, Costa Rica, 1984.
23. Vargas, A. *Evaluación del Forraje de Poró (Erythrina cocleata) como Suplemento Proteico para Toretes en Pastoreo*; Master's Thesis; CATIE: Turrialba, Costa Rica, 1987.
24. Jackson, A.H.; Ludgate, P.; Mavraganis, V.; Redha, F. *Allertonia* **1982**, *3*, 47.
25. Marten, G.C.; Jordan, R.M.; Hovin, A.W. *Crop Sci.* **1981**, *21*, 295-298.
26. Víquez, E.; Payne, L.D.; Sanchéz, G.A. *El Huerto Latinoamericano de Arboles Fijadores Nitrógeno*; First National Forestry Congress: San José, Costa Rica, November 11-14, 1986.
27. Hargreaves, R.T.; Johnson, R.D.; Millington, D.S.; Mondal, M.H.; Beavers, W.; Becker, L.; Young, C.; Rinehart Jr., K.L. *Lloydia* **1974**, 37, 569-580.
28. Barakat, I.; Jackson, A.H.; Abdullah, M.I. *Lloydia* **1977**, *40*, 471-475.

29. Redha, F.J.M.*Chromatographic and Spectroscopic Studies of Erythrina Alkaloids*; Ph.D. Dissertation; University of Wales: Cardiff, Great Britain, 1983.
30. Bhakuni, D.S.; Uprety, H.; Widdowson, D. A. *Phytochem.* **1976**, *15*, 739-741.
31. Bick, I.R.C.; Doebel, K.; Taylor, W.I.; Todd, A.R. *J. Chem. Soc.* **1953**, *1953*, 692-695.
32. Bick, I.R.C.; Taylor, W.I.; Todd, A.R. *J. Chem. Soc.* **1953**, *1953*, 695-700.
33. Keeler, R.F.; Binns, W. *Phytochem.* **1971**, *10*, 1765-1769.
34. Blua, M. J.; Hanscom, Z. *J. Chem. Ecol.* **1986**, *12*, 1449-1458.
35. Krukoff, B.A.; Barneby, R.C. *Lloydia* **1974**, *37*, 332-459.
36. Cranmer, M.F.; Turner, B.L. *Evolution* **1967**, *21*, 508-517.
37. Ito, K.; Furukawa H.; Tanaka, H. *Chem. Pharm. Bull.* **1971**, *19*, 1509-1511.
38. Ito, K.; Haruna, M.; Jinno, Y.; Furukawa, H. *Chem. Pharm. Bull.* **1976**, *24*, 52-55.
39. Chunchatprasert, S. *Chemical and Spectroscopic Studies of Erythrina Alkaloids*; Ph.D. Thesis; University of Wales: Cardiff, Great Britain, 1982.
40. Romeo, J.T.; Bell, E.A. *Lloydia* **1974**, *37*, 543-568.
41. Gonzales, V.; Brewbaker, J.L.; Hamill, D.E. *Crop Science* **1967**, *7*, 140-143.
42. Dyke, S.F.; Quessy, S.N. *The Alkaloids*; Academic Press: New York, NY, 1981; Vol. XVIII, 1.

RECEIVED February 24, 1992

Chapter 8

Developing Stereoselective High-Performance Liquid Chromatographic Assays for Pharmacokinetic Studies

Irving W. Wainer

Division of Pharmacokinetics, Department of Oncology, McGill University, Montreal, Quebec H4G 1A4, Canada

Enantioselective chromatography using HPLC chiral stationary phases (HPLC-CSPs) is rapidly becoming a stand procedure in analytical laboratories. This technique has been extensively used with pharmacologically acitve chiral compounds, although most of the reported separations are for the bulk drug substance. The application of HPLC-CSPs to pharmacokinetic and metabolic studies presents the analyst with addition problems often caused by interferrences from the matrix or from metabolites. One method to overcome these difficulties is coupled achiral/chiral chromatography which is discussed below.

In the past few years there has been an increased interest in the pharmacological fate of chiral substances and a rapid growth in the study of the pharmacokinetic and metabolic disposition of stereoisomeric drugs. These development have been primarily due to the development and commercial availability of chiral stationary phases for high performance liquid chromatography (HPLC-CSP's). These phases have formed the backbone of numerous analytical methods capable of determining the *in vivo* concentrations of the various stereoisomers.

Since the introduction in 1981 of the (R)-N-(3,5-dinitrobenzoyl)phenylglycine CSP developed by W.H. Pirkle [1], the number of commercially available HPLC-CSPs has grown to over 55. This wide variety of HPLC-CSP's has expanded the ability of the analytical chemist to develop the necessary assays to follow the *in vivo* fate of chiral drugs. It has also made it easier; for when there was only one or two HPLC-CSP's on the market it was often necessary to extensively alter the solutes to fit the requirements of the chiral recognition mechanisms operating on the CSP. This is no longer the case. With the large number of available CSPs, a column can usually be picked to fit the properties of the solutes, eliminating the necessity for derivatization, and the requirements of the assay. This situation is illustrated by the development of two assays for the determination of the serum concentrations of the enantiomers of propranalol.

0097–6156/92/0512–0100$06.00/0

Serum Concentrations of Propranolol Enantiomers

The first assay for the serum concentrations of the propranalol enantiomers using an HPLC-CSP was accomplished in 1983 using the (R)-N-(3,5-dinitrobenzoyl)phenyl-glycine CSP, DNPG-CSP, [2]. At that time, this was the only commercially available HPLC-CSP. In order to achieve an enantioselective separation on this CSP, the solute should contain one or more of the following functional groups: a π-acid or a π-base; a hydrogen bond donor and/or a hydrogen bond acceptor; an amide dipole [3]. The solute molecule should also not contain functional groups which are strongly cationic, i.e. a free amine moiety, or anionic, i.e. a carboxylic acid moiety [3].

From the molecular structure of propranalol, it is clear that the substance had to be derivatized to eliminate the undesirable interactions caused by the free amine and to create additional positive interaction sites with the CSP. This was accomplished by reacting propranolol with phosgene to form an oxazoladone (Figure 1) which could then be chromatographed giving the separation illustrated in Figure 2.

Since the assay had to be based upon the requirements of the CSP, the resulting procedures required the initial extraction of the target drugs from plasma using diethyl ether, addition of phosgene (12.5% solution in toluene), collection and evaporation of the ether layer, redissolution of the resulting solid in methylene, and injection onto the CSP. The observed retention times of the propranolol enantiomers were 4,839 and 5259 sec, Figure 2, making the assay both lengthy and complicated, especially since it employed a rather toxic derivatizing agent.

Subsequent to the development and application of the assay utilizing the DNPG-CSP, a series of HPLC-CSPs based upon derivatized cellulose were developed; including a phase composed of cellulose tris(3,5-dimethylphenylcarbamate), the OD-CSP [3]. Using the OD-CSP, the enantiomers of propranalol could be enantioselectively resolved directly on the column without prior derivatization. This allowed for the development of a new assay which required only three steps: 1) the addition of sodium hydroxide to the serum; 2) extraction of the propranolol into an organic solvent; 3) injection of that extract onto the OD-CSP [4]. The resulting chromatograms are given in Figure 3. Thus the technological advancement of going from one column, the DNPG-CSP to the OD-CSP, reduced both the preparation and analysis times as well as producing an assay which was more "user friendly".

The Need for Coupled Achiral/Chiral Chromatographic Systems

As illustrated by the substitution of the OD-CSP for the DNPG-CSP, technological advances in the development of new CSP's has permitted the development of direct and simple assays. The drive of the analytical chemist is always to develop assays which consume the minimum amount of supplies (i.e. organic solvents, etc.) and technician time (i.e. extractions, derivatizations and length of chromatographic run). This had been made possible by the development of the wide variety of HPLC-CSPs. However, while the HPLC-CSPs are able stereochemically resolve the individual isomers of a chiral molecule, and at this stage it is safe to say that there are very few enantiomers which cannot be separated using this technology, they often cannot

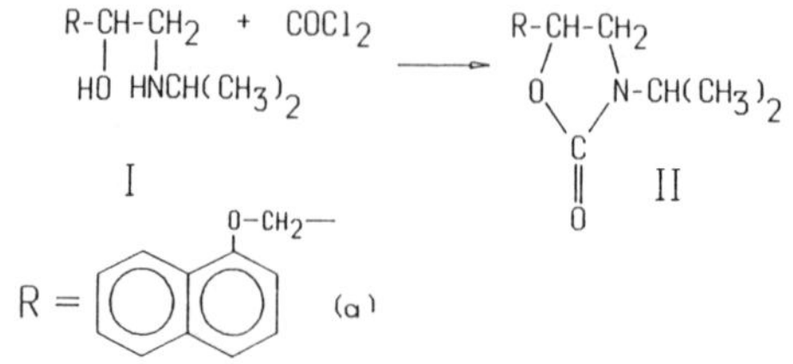

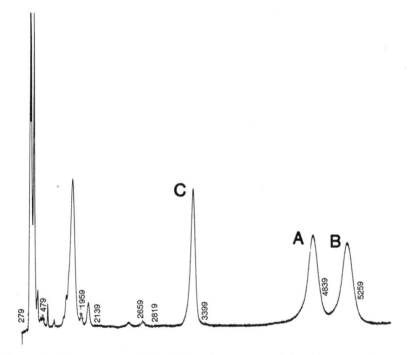

Figure 1. Synthesis of the oxazolidone of propranolol. See Reference 2 for experimental details.

Figure 2. Chromatogram of whole blood extract containing 50 ng racemic propranolol per ml. Peaks: A = oxazolidone corresponding to (S)-propranolol; B = oxazolidone corresponding to (R)-propranolol; C = internal standard. For chromatographic conditions see Reference 2.

differentiate between structurally related compounds. This is not a problem when you are dealing with the analysis of a chiral compound as a bulk drug substance or in a pharmaceutical formulation. However, it is a problem when you are dealing with drugs and their metabolites. This problem is illustrated by the example of verapamil (VER) and one of its major metabolites, norverapamil (NORVER).

Determination of Verapamil and Norverapamil Enantiomers in Serum - First Approach. Using a CSP based upon immobilized α_1-acid glycoprotein (AGP-CSP), VER and NORVER can be readily separated without derivatization, Figure 4A and 4B, respectively. However, when you chromatograph the parent, VER and the metabolite NORVER together, which would be the situation after administration of VER to a living organism, what you would find is that the compounds overlap as illustrated in Figure 4C. Thus, the AGP-CSP was able to stereochemically resolve the respective enantiomers of VER and NORVER, but unable to separate parent from metabolite.

One approach to overcoming this problem of structural selectivity and to maximize the utility of the CSP's enantioselectivity is to use coupled achiral/chiral chromatography. In this approach an achiral column is used to separate the target compound or compounds from the biological matrix, as well as from each other. The eluents containing the target compound or compounds are then directed onto the CSP where they are stereochemically resolved. A diagram of one type of coupled achiral/chiral chromatographic system is presented in Figure 5.

The system presented in Figure 5 has been used to solve the problems presented by VER and to analyze the enantiomeric compositions of both VER and NORVER in plasma [5]. In this assay, the achiral column contained a shielded hydrophobic phase (Hisep) which was able to separate VER from NORVER and both compounds from the other components in the plasma matrix. The eluents containing VER and NORVER were selectively transferred to the AGP-CSP where the enantiomers were stereochemically resolved and the enantiomeric composition determined. The resultant chromatograms from the AGP-CSP are presented in Figure 6.

This assay has been validated and used extensively in single dose studies. However, while this is a successful application of the achiral/chiral coupled column system depicted in Figure 5, there are a number of problems which plague this type of coupled column system. For example, a single assay requires 2 injections -the first is used to quantitate the total drug on the achiral system and the eluent passes from the achiral column to the detector and then to waste; the second injection is used to provide the eluent which is switched to the CSP for enantiomeric analysis.

This approach is not only cumbersome and time consuming, it is often insensitive on the chiral end of the analysis due to loss of compound during the switching process and to band broadening. While it is often impossible to avoid the type of coupled achiral/ chiral chromatography required for the separation of VER and NORVER on the AGP-CSP, the continuing development of new HPLC-CSPs often presents new possibilities. This can also be illustrated by the development of a new assay for the enantiomeric composition of VER and NORVER in plasma.

Determination of Verapamil and Norverapamil Enantiomers in Serum - Second Approach. Okamoto, *et al.* [6] have recently reported the development of a new series of HPLC-CSPs based upon amylose, and in particular, a 3,5 dimethylphe-

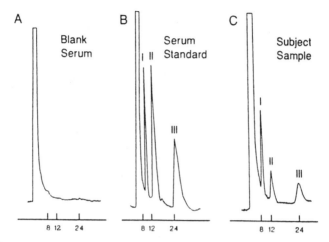

Figure 3. Representative chromatograms for: A = an extracted blank serum sample; B = serum sample spiked with racemic propranolol (150 ng per ml); C = serum sample from a volunteer subject 12 h after ingestion of a 160-mg dose of racemic propranolol. See Reference 4 for chromatographic conditions.

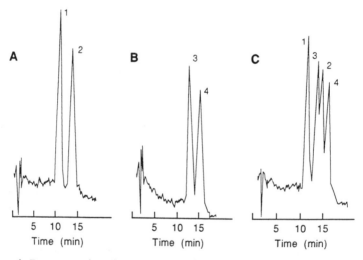

Figure 4. Representative chromatograms from the chromatography of racemic verapamil (VER) and racemic norverapamil (NORVER) on the AGP-CSP. A = racemic VER; B = racemic NORVER; C = 50:50 mixture, racemic VER and racemic NORVER. Peaks: 1 = (R)-VER; 2 = (S)-VER; 3 = (R)-NORVER; 4 = (S)-NORVER. See Reference 6 for chromatographic conditions.

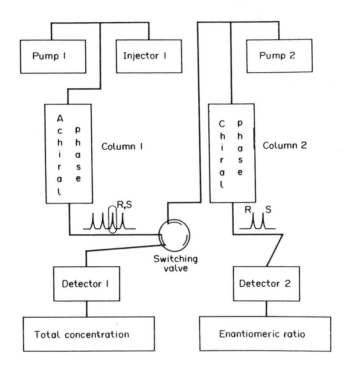

$$[\text{R-isomer}] = [\text{Total}] \cdot \text{R/S}$$
$$[\text{S-isomer}] = [\text{Total}] - [\text{R}]$$

Figure 5. Schematic representation of a Coupled Achiral/Chiral Chromatographic System.

nylcarbamate-derivatized amylose coated on aminopropyl silica, AD-CSP. The AD-CSP is able to achieve the enantioselective resolution of underivatized VER and NORVER without overlap has allowed for the direct separation of verapamil and norverapamil without the necessity of column switching.

However, when this column was applied to chronic dosing studies, it became evident that there were additional interferences with the chromatography of VER and NORVER from a variety of other VER metabolites which are listed in Table 1, [7]. As you can see from this table, on the AD-CSP, (R)-VER and metabolite D617 overlapped as did (S)-NORVER and metabolite PR25.

The problem presented by the coelution of (R)-VER and D617 and (S)-NORVER and PR25 was solved by another form of coupled column chromatography in which an achiral column containing a LiChrocart diol silica was placed in front of the AD-CSP and in series with the CSP [7]. This changed the retention times of the metabolites as well as VER and NORVER, Table 1, and allowed for the total separation of all the compounds. The resulting chromatogram is presented in Figure 7.

TABLE 1.

RETENTION TIMES OF VERAPAMIL AND ITS METABOLITES ON AN AD-CSP WITHOUT (A) AND WITH (B) A LICHROCART DIOL COLUMN COUPLED IN SERIES. Chromatographic conditions: mobile phase, hexane: propanol-2:ethanol (85:7.5:7.5, v/v/v) containing 1.0% triethylamine; flow rate, 1.0 ml/min; excitation, 272 nm; emission, 317 nm; temperature, ambient. See reference [7] for further details

Compound	Retention time (min)	
	Without DIOL	With DIOL
(S)-Verapamil	6.7	7.4
(R)-Verapamil	7.7	8.4
(S)-Norverapamil	10.5	12.2
(R)-Norverapamil	11.7	13.4
D617	7.4	17.0
PR25	10.7	32.8
PR22 (1st isomer)	15.5	18.6
PR22 (2nd isomer)	18.2	21.2

Sequential Achiral/Chiral Chromatography

When there are many metabolites to deal with or when the biological matrix is extremely complicated the analytical chemist may be forced into another form of coupled column chromatography, sequential achiral/chiral chromatography. While this is in fact the easiest of the coupled column chromatographic methods to develop it is also the one which is the most time consuming. The approach is as follows: the compound or compounds of interest are separated first on an achiral column; the

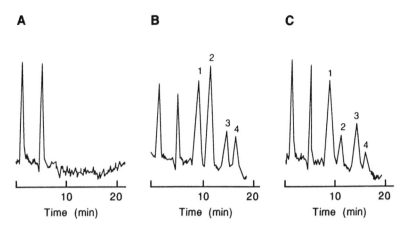

Figure 6. Representative chromatograms from the chromatography of verapamil (VER) and norverapamil (NORVER) on the AGP-CSP coupled to the Hisep HPLC column. A = serum blank; B = serum spiked with 100 ng/ml each of racemic VER and NORVER; C = serum sample taken 12 h after the administration of a 240-mg sustained release dose of racemic VER. Peaks: 1 = (R)-VER; 2 = (S)-VER; 3 = (R)-NORVER; 4 = (S)-NORVER. See Reference 6 for chromatographic conditions.

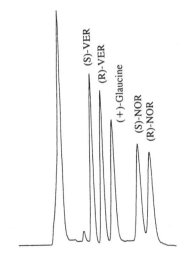

Figure 7. Representative chromatograms from the chromatography of verapamil (VER) and norverapamil (NORVER) in plasma on an HPLC system with a DIOL column (5 cm x 4.0 mm I.D.) coupled in line to an AD-CSP (25 cm x 4.6 mm I.D.). See Reference 7 for chromatographic conditions.

eluent containing these compounds are collected and concentrated; then reinjected onto the CSP. This approach is illustrated in Figure 8.

The collection of the eluents and their concentration all take place off-line requiring additional labor in preparing the compounds for the second injection. A mixture which contains four target compounds will require a single injection to separate the compounds and four additional injections of each one of the isolated compounds; thus, 100 samples requires 500 injections.

While this method is time consuming, it often the only way to solve the problem. This situation is illustrated by the assay for the enantiomeric composition of hydroxychloroquine and its major metabolites in plasma [8]. In this approach, hydroxychloroquine and its metabolites desethylchloroquine, desethylhydroxychloroquine and bidesethylchloroquine were first separated on a achiral column containing a cyano-bonded stationary phase. The eluents were collected and concentrated on a Speed-Vac evaporator and reconstituted in the mobile phase used on the AGP-CSP, the CSP used in this assay. Then each one of the compounds was injected on to the AGP-CSP. The chromatographic parameters for these separations are presented in Table 2.

TABLE 2.

CHROMATOGRAPHIC PARAMETERS OF HYDROXYCHLOROQUINE (HCQ), BIDESETHYLCHLOROQUINE (BDCQ), DESETHYLHYDROXYCHLOROQUINE (DHCQ) AND DESETHYLCHLOROQUINE (DCQ) ON AN ACHIRAL CYANO-BONDED PHASE AND THE AGP-CSP. See reference [8] for further details

---

Compound	Cyano column k'^a	AGP-CSP k'^b	α^c	$R_{RS}{}^d$
BDCQ	1.61	12.11	1.25	1.29
DHCQ	2.56	10.19	1.32	1.39
DCQ	3.93	11.44	1.39	1.97
HCQ	5.33	8.67	1.39	2.08

---

[a] Capacity factor; [b] Capacity factor first eluted enantiomer;
[c] Enantioselectivity factor; [d] Enantiomeric resolution

The sequential achiral/chiral assay for the enantiomeric concentrations of hydroxychloroquine and its enantiomers in plasma has been validated and is now in use in pharmacokinetic and metabolic studies.

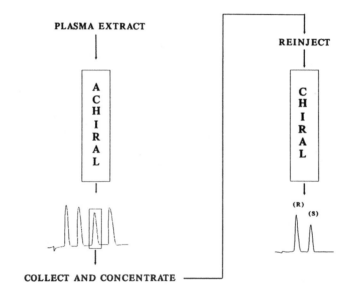

Figure 8. Schematic representation of a Sequentially Coupled Achiral/Chiral Chromatographic System.

Conclusion

While the development of pharmacokinetic and metabolic studies of the enantiomeric compounds in biological fluids may at first seem difficult and time consuming, it is only because we are currently in a transition from the use of HPLC-CSPs solely for the separation of bulk drugs to their routine application in the bioanalytical laboratory. Within a few years this will change and these phases will be commonly found in the pharmacokinetic and bioanalytical laboratories. Along with the standard use of HPLC-CSP will come column switching and similar techniques which are a normal part of application of these CSPs. On the other hand, the development of improved CSPs and the progress in enantioselective separations utilizing newer technologies such as capillary electrophoresis may make these approaches obsolete. What ever happens, there is an exciting and challenging future ahead for the separation of enantiomeric compounds and the understanding of their pharmacokinetic and metabolic fate.

Literature Cited

[1]   Pirkle, W.H., Finn, J.M., Schriner, J.L., Hamper, B.C. *J. Am. Chem. Soc.* **1981**, *103*, 3964-3966.

[2]   Wainer, I.W., Doyle, T.D., Donn, K.H., Powell, J.R. *J. Chromatogr.* **1984**, *306*, 405-411.

[3]   Wainer, I.W. *A Practical Guide to the Selection and Use of HPLC Chiral Stationary Phases*, J.T. Baker Inc., Phillipsburg, NJ, 1988.

[4]   Straka, R.J., Lalonde, R.L., Wainer, I.W. *Pharm. Res.* **1988**, *5*, 187-189.

[5]   Chu, Y.-C., Wainer, I.W. *J. Chromatogr.* **1989**, *497*, 191-200.

[6]   Okamoto, Y., Aburatani, R., Fukumoto, T., Hatada, K. *Chem. Lett.* **1987**, 411-414.

[7]   Shibukawa, A., Wainer, I.W. *J. Chromatogr.* **1992**, *574*, 85-92.

[8]   Iredale, J., Wainer, I.W. *J. Chromatogr.* **1992**, *573*, 253-258.

RECEIVED May 5, 1992

Chapter 9

Direct Enantiomeric Separation and Analysis of Some Aromatase Inhibitors on Cellulose-Based Chiral Stationary Phases

Hassan Y. Aboul-Enein

Drug Development Laboratory, Radionuclide and Cyclotron Operations Department, MBC–03, King Faisal Specialist Hospital and Research Centre, P.O. Box 3354, Riyadh 11211, Kingdom of Saudi Arabia

The stereoisomeric composition of pharmaceuticals is rapidly becoming one of the key issues in the development of new drugs. Several chiral stationary phases (CSP's) are now available to effect direct separation and analyses of drug enenatiomers and racemates. Cellulose CSP's are one of these commonly employed phases which have been successfully used in separation, enantiomeric purity determination and analysis of several aromatase inhibitors belonging to the glutarimide group namely aminogluthethimide (1), pyridogluthethimide (2), cyclohexylaminogluthethimide(3). A comprehensive presentation of baseline resolution of these drugs is presented.

Aminogluthethimide (AG, 1) was first used as anticonvulsant, but was observed later to cause adrenal insufficiency which led to its withdrawal from the market. The drug inhibits several enzymes in the pathway of steroidogenesis, mainly the desmolase enzyme system which is responsible for the cholesterol side chain cleavage, i.e., conversion of cholesterol to pregnenolone (1), and the aromatase enzyme system which is responsible for the aromatization of androstene-3,17-dione and testosterone to esterone and estradiol, respectively (2). Thus, AG can interfere with the biosynthesis of mineralocorticoids, glucocorticoids and sex hormones and effectively perform a physiological adrenalectomy. Aminogluthethimide has been used as a racemic mixture in the treatment of metastatic estrogen-dependent breast cancer as a reversible non-steroidal aromatase inhibitor (3,4).

However, due to its inhibitory effect on desmolase, corticosteroid production is depleted, and consequently patients receiving AG

0097–6156/92/0512–0111$06.00/0
© 1992 American Chemical Society

require hydrocortisone replacement therapy to prevent the reflex rise in adrenocorticotropic hormone (ACTH) which can counteract the initial blockade of desmolase. Furthermore, several side effects were reported during AG administration e.g., sedations, CNS depression, ataxia and neurotoxic effects (5).

It is of interest to mention that the (+)-R enantiomer of AG is 30 times more potent than the (-)-S enantiomer in aromatase inhibition, whereas the (+)-R enantiomer is 15 times less potent than the (-)-S isomer in inhibiting desmolase (6,7).

Several analogs of AG were developed through substitution of the ethyl group by other bulkier groups or by replacing the p-aminophenyl group by other basic functionalities in order to design specific and selective aromatase inhibitors with little or no desmolase inhibitory activity. Subsequently, the need for hydrocortisone supplement administration will not be required.

Foster et al.(8) reported the synthesis of pyridogluthethimide (PG, 2) which is considered a bioisoster of AG. Pyridogluthethimide exhibited a strong, selective and competitive inhibitory activity against aromatase while it did not inhibit the desmolase enzyme system. Pyridogluthethimide is administered as a racemic mixture, although it is reported that the (+)-R-enantiomer is more active than the (-)-S-enantiomer (9).

Cyclohexylaminogluthethimide (ChAG, 3), another active analog of AG, first synthesized by Hartmann et al. (10) showed a strong and selective inhibitory activity (123-fold in vitro and 10-fold in vivo and compared to AG) against aromatase, while the inhibition against the desmolase enzyme system was greatly reduced. Hartmann et al. (11) also reported that the (+)-S-enantiomer proved to be responsible for the aromatase inhibition and is 30 times more active than the (-)-R-enantiomer. Furthermore, (+)-S-ChAG showed no CNS depressive activity and possessed a weaker effect regarding the inhibition of the desmolase compared to its (-)-R-enantiomer. The chemical structures of these drugs are shown in Figure 1.

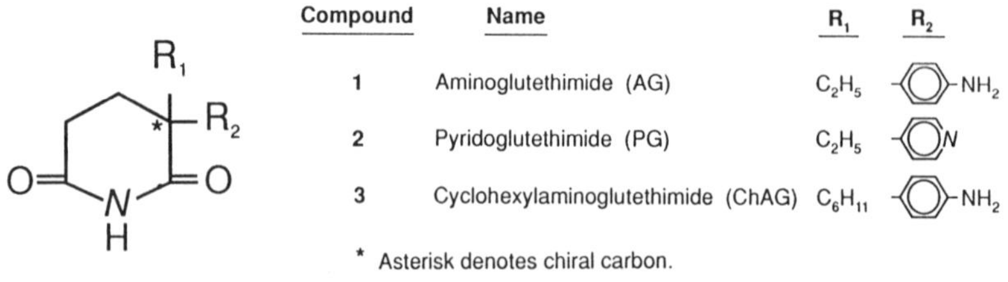

Compound	Name	R_1	R_2
1	Aminoglutethimide (AG)	C_2H_5	⬡-NH$_2$
2	Pyridoglutethimide (PG)	C_2H_5	⬡N
3	Cyclohexylaminoglutethimide (ChAG)	C_6H_{11}	⬡-NH$_2$

* Asterisk denotes chiral carbon.

Figure 1. The chemical structures of various glutarimide aromatase inhibitors.

Thus, from the previous introduction, it would be clinically more useful to administer these aromatase inhibitors as pure enantiomers namely: (+)-R-aminogluthethimide (+)-R-pyridogluthethimide and (+)-S-cyclohexylaminogluthethimide rather than other racemate mixtures. This will reduce the side effects encountered by the other enantiomers and also allow the use of appropriate smaller doses, thus achieving more effective therapy.

This paper describes a method for the direct separation of the racemates of these aromatase glutarimide inhibitors to their corresponding enantiomers using commercially available cellulose-based chiral stationary phases (CSP's) namely Chiralcel OD (cellulose tris-3,5-dimethylphenyl carbamate) and/or Chiralcel OJ (cellulose tris-4-methyl benzoate). The optimum chromatographic conditions for the separation of these drugs on these CSP's are also studied.

EXPERIMENTAL

Apparatus

The Waters Liquid Chromatography System used (Waters Association, Milford, MA) consisted of a Model M-45 pump, a U6K injector, and a Lambda-Max Model 481 LC spectrophotometer detector operated at 257 nm. Chiralcel OD and Chiralcel OJ coated on silica gel with particle size of 10um were used as stationary phases (25cm x 0.46cm i. d., Daicel Chemical Industries, Tokyo, Japan).

Chemicals

Racemic aminoglutethimide (±AG) (Lot No. 800383), (+)-R-aminoglutethimide (+AG) (CGS-2396) were supplied by Ciba-Geigy (Basle, Switzerland). Racemic pyridoglutethimide (±PG), and its corresponding (+)-R and (-)-S-enantiomers were kingly supplied by Dr. M. Jarman of the Institute of Cancer Research, Sutton, Surrey, U.K. Racemic cyclohexylaminogluthethimide (±ChAG) and its corresponding (-)-R-enantiomer were kindly supplied by Professor R.W. Hartmann, University of Saarland, Saarbrücken, Germany. HPLC grade hexane and HPLC grade 2-propanol were purchased from Fisher Scientific, Fairlawn, NJ.

Chromatographic Conditions

The resolution of AG, PG and ChAG were obtained on the Chiralcel OD column using hexane: 2-propanol as a mobile phase in proportions shown in Figure 2, 3 and 4, respectively. Racemic AG and PG could also be resolved successfully on an OJ column using hexane: 2-propanol (50:50) as a mobile phase as shown in Figures 5 and 6, respectively. Temperature was maintained at 23°C throughout the experiments.

Determination of Enantiomeric Elution Order

The enantiomeric elution order was determined by chromatographing the separate individual enantiomers under similar conditions. Thus, in the case of AG on both the Chiralcel OD and OJ columns the peak that eluted first was identified as (-)-S-AG while the second peak was identified as (+)-R-AG

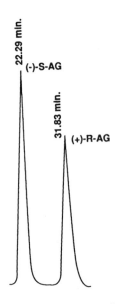

Figure 2.   LC separation of racemic AG (<u>1</u>). Column: Chiralcel
            OD (250 X 4.6mm, I. D.); mobile phase, hexane:
            2-propanol (60:40); flow rate: 0.7 ml/min;
            chart speed: 0.25 cm/min; temperature: 23°C;
            detector: UV 257nm; sensitivity 0.01 AUFS; sample
            amount: 2.5 nmol.

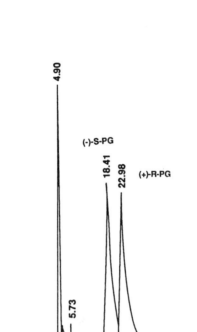

Figure 3. LC separation of racemic PG (2). Chromatographic
conditions were the same as in Figure 2, except
mobile phase, hexane: 2-propanol (65:35); sample
amount 2.8 nmol.

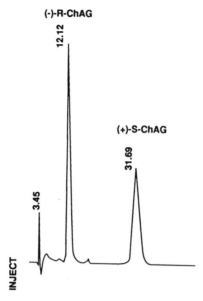

Figure 4.   LC separation of racemic ChAG (<u>3</u>). Chromato-
            graphic conditions were the same as in Figure 2,
            except mobile phase, hexane: 2-propanol (50:50);
            flow rate, 1.0 ml/min; chart speed, 0.2 cm/min;
            sample amount, 2.5 nmol.

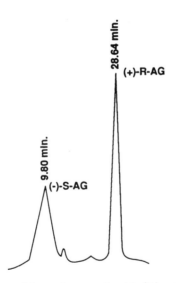

Figure 5.   LC separation of racemic AG (<u>1</u>). Column: Chiralcel
            OJ (250 X 4.6mm, I. D.); mobile phase, hexane:
            2-propanol (50:50). Other chromatographic
            conditions were the same as in Figure 2, except
            sample amount: 3 nmol.

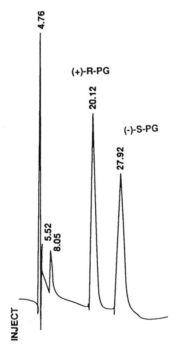

Figure 6.   LC separation of racemic PG (2). Column:
Chiralcel OJ (250 X 4.6 mm, I. D.); other
chromatographic conditions were the same as in
Figure 5, except chart speed: 0.2 cm/min;
sample amount: 2.8 nmol.

In the case of the resolution of the PG enantiomers on
the Chiralcel OD column, the peak that eluted with a lower capacity
factor was the (-)-S-PG and the peak that eluted with a higher
capacity factor was identified as (+)-R-PG.   However,
the elution order obtained for the resolution of (±) PG on OJ
column was reversed.

In the case of ChAG, it was found that the peak eluted with a
lower capacity factor was identified as (-)-R-ChAG and the peak
that eluted with a higher capacity factor was identified as
(+)-S-ChAG.

RESULTS AND DISCUSSION

The stereochemical composition of pharmaceuticals is becoming an
important issue not only in the course of drug development, but
also drug approval and clinical use. At present, the pharmaceutical
industry is rapidly moving towards precisely targeted, zero-risk
drug therapies. The fast development of chiral stationary phases
in recent years had demonstrated the versatility and convenience
of the technique in chiral separation of drug racemates to their
corresponding enantiomers.

Cellulose derivatives function as chiral absorbents and exhibit
good performance comparable to other phases. Cellulose CSP's
belong to the helical polymer phases, namely the cellulose esters
and carbamates whose chirality arises from helicity.

Various types of cellulose and derivatives carrying various
substituents have been developed (12). The cellulose derived
Chiralcel OD column has been successfully used to directly separate
several β-adrenergic blockers (13-18). Chiralcel OJ has also
been used to separate drugs chemically related to the glutarimide
ring system, e.g., thalidomide (19) and mephobarbital (20). In
this study Chiralcel OD and Chiralcel OJ were successfully used
in the resolution of the racemic mixture of AG, PG, and ChAG.
Optimization of the separation of these drugs was achieved using
different concentrations of 2-propanol in hexane as a mobile
phase (see Table I).

Since good stereochemical resolution was obtained for these drugs,
this method is currently applied for preparative separation of
the pharmacologically active enantiomers (eutomers) in large
quantities. Furthermore, the method could be used for enantiomeric
and optical purity determination of these drugs in bulk form and
pharmaceutical formulations. The described method has an
advantage of being fast and requires no derivatization. Work is
in progress in this laboratory to validate the method for
analysis of these drugs in biological fluids.

In conclusion, methods for the direct enantiomeric separation of
various non-steroidal glutarimide aromatase inhibitors have been
developed using cellulose-based CSP's (Chiralcel OD and
OJ columns).

Table I. Optimized parameters of capacity factor (k'), stereochemical separation factor (α) and stereochemical resolution (R) of AG, PG and ChAG on Chiralcel OD and Chiralcel OJ columns

Compound	Solvent	Column	k'_1	k'_2	α	R
AG	A	OD	6.61	9.96	1.51	8.87
AG	B	OJ	2.35	8.78	3.74	10.34
PG	C	OD	2.76	3.69	1.34	0.96
PG	B	OJ	3.15	4.74	1.50	1.56
ChAG	B	OD	2.51	8.18	3.26	4.89

Chromatographic Conditions

Solvent system A = hexane: 2-propanol (60:40)
Solvent system B = hexane: 2-propanol (50:50)
Solvent system C = hexane: 2-propanol (65:35)
Detector : UV257nm; temperature 23°C.
k'_1 = 1st eluted peak
k'_2 = 2nd eluted peak
α = stereochemical factor
R = stereochemical resolution

ACKNOWLEDGEMENTS

The author would like to thank the administration of King Faisal Specialist Hospital and Research Centre for their continuous support to the Drug Development Research Program. This investigation was supported financially under Project No. 88-0015 by King Faisal Specialist Hospital & Research Centre. The author wishes to thank Dr. K. Scheibli, Ciba-Geigy, Basle, Switzerland, Dr. M. Jarman, Institute of Cancer Research, Sutton, Surrey, U.K. and Dr. R.W. Hartmann, Institute of Pharmaceutical Chemistry, University of Saarland, Saarbrucken, Germany for providing the samples of drugs used in this study.

LITERATURE CITED:

1. Lonning, P. E. and Kvinnsland, S. (1988). Drugs 35, 685-710.

2. Dexter, R. N., Fishman, L. M., Ney, R. L. and Liddle, J. (1967), J. Clin. Endocrinol 27, 374-480.

3. Aboul-Enein, H. Y., (1988). Drug Design & Delivery, 2, 221-226, and references cited therein.

4. Katzung, B. G., (1989). Basic & Clinical Pharmacology 4th Ed. Appleton & Lange, East Norfolk, CT (USA), p. 480.

5. Salhanick, H. A., (1982). Cancer Res. (Suppl.), 42, 3315S-3321S.

6.   Uzgiris, V. I., Whipple, C. A. and Salhanick, H. A.,
     Salhanick, (1977). Endocrinology, 101, 89-92.

7.   Finch, N., Dziemian, R., Cohen, J. and Steinetz,
     B. G., (1975). Experientia, 31, 1002-1003.

8.   Foster, A. B. Jarman, M., Leung, C. S., Rowlands,
     M. G., Taylor, G. N., Plerey, P. G. and Sampson, P.,
     (1985). J. Med. Chem., 28, 200-204.

9.   Clissold, D. W., Jarman, M., Mann, J., McCague, R.,
     Neidle, S., Rowlands, M. G., Thickitt, C. P. and
     Webester, G., (1989). J. C. S. Perkin Trans. I.
     196-198.

10. Hartmann, R. W., Batzl, C., Mannschreck, A. and Seydel
     J. K. (1989),. Trends in Medicinal Chemistry '88;
     van der Goot, H., Domany, G., Pallos, L., Timmermann,
     T. (Eds.), Elsevier Science Publishers, Amsterdam, The
     Netherlands, pp 821-838.

11. Hartmann, R. W., Batzl, C., Mannschreck, A., Pongratsz,
     T., (1990). Chirality and Biological Activity;
     Holmstedt, B., Frand, H., Testa, B. (Eds).; Alan B.
     Liss: NY, USA, pp 185-190.

12. Aboul-Enein, H. Y., Islam, M. R., (1990). J. Liq.
     Chromatogr., 13, 485-492., and references cited therein.

13. Okamoto, Y., Kawashima, M., Aburatani, R., Hatada, K.,
     Nishiyama, T., Matsuda, M., (1986), Chem. Lett.,
     1237-1240.

14. Aboul-Enein, H. Y. and Islam, M. R., (1990). J.
     Chromatogr., 511, 109-114.

15. Aboul-Enein, H. Y. and Islam, M. R., (1989). Chirality,
     1, 301-304.

16. Aboul-Enein, H. Y. and Islam, M. R., (1990), Anal.
     Lett. 23, 83-91.

17. Aboul-Enein, H. Y. and Islam, M. R., (1990), Anal.
     Lett. 23, 973-980.

18. Krstulovic, A.M., Fouchet, M. H., Burke, J. T., Gillet,
     G. Durand, A., (1988). J. Chromatogr., 452, 477-483.

19. Aboul-Enein, H. Y. and Islam, M. R., (1991). J. Liq.
     Chromatogr., 14, 667-673.

20. Okamoto, Y., Aburatani, R., Hotada, K., (1987). J.
     Chromatogr., 389, 95-102.

RECEIVED February 24, 1992

Chapter 10

Chromatographic Analysis of Host-Cell Protein Impurities in Pharmaceuticals Derived from Recombinant DNA

Donald O. O'Keefe and Mark L. Will

Department of Analytical Research, Merck Sharp and Dohme Research Laboratories, Rahway, NJ 07065

The purity of recombinant protein therapeutics is of paramount concern. Impurities in these drugs can be potentially unsafe and can include endotoxin, DNA, and host-cell proteins. The ability to detect, identify, and quantitate these impurities is an ongoing challenge for the bioanalyst. In this chapter, we provide an overview of the current chromatographic methods used, including gel electrophoresis, to analyze host-cell protein impurities. The advantages and disadvantages of each method are presented and demonstrate that any technique by itself is insufficient, but several techniques together provide a more complete and reliable protein impurity analysis.

The arrival of recombinant DNA technology has provided the pharmaceutical industry with a unique source of therapeutic proteins. This advent in technology has also generated new problems for the pharmaceutical analyst. Among these problems are the detection, quantitation, and identification of drug impurities. There are four basic sources of impurities in protein pharmaceuticals: 1) the fermentation medium 2) the host cells, 3) the purification process, and 4) the protein drug itself. Medium-related impurities include additives used to supplement host-cell growth. Certain cell lines, notably those of mammalian origin, are routinely supplemented with protein additives. Host-related impurities include endotoxin, DNA, and proteins. Impurities derived from the process can include Protein A and any monoclonal antibodies used in affinity chromatography. Drug-related impurities include aggregates, degradates, structural isomers, and compositional isomers. The limited spectrum of impurities presented here makes the complex analysis apparent. Our intention in this chapter is to focus on one type of impurity, that of host-cell proteins.

The level of host-cell protein impurities in an unprocessed recombinant therapeutic depends on the cell line used for production. Currently, recombinant

0097–6156/92/0512–0121$06.00/0

proteins are produced in either *Escherichia coli,* yeast or mammalian cells. Early process samples of recombinant products derived from either yeast or mammalian systems are often highly enriched in the drug because the protein is usually secreted into the extracellular medium. Proteins produced in *E. coli* are not secreted and are only obtained after the cell is lysed. Bacterial cells have been estimated to contain a minimum of 1500 different proteins. Any number of these cellular proteins are potential impurities. This underscores the difficulty faced in both the processing and bioanalytical areas.

Guidelines for the analysis of recombinant protein therapeutics for host-cell protein impurities are provided by the Food and Drug Administration (FDA)(*1*). The tolerable impurity level is reflected by the drug dose and the mode of administration. Initial analysis focuses on detecting protein impurities. In contrast to more traditional drugs, proteins are large and often immunogenic in trace amounts. An immunogenic impurity could act as an adjuvant when administered along with the drug. This can lead to an immunological response towards the drug itself rendering it ineffective. Highly sensitive assays are required to determine both aggregate and individual levels of impurities. Often sensitivities down to ppm (parts per million by weight) or lower are required (*2*).

The identity of protein impurities provides valuable information on several fronts. Firstly, identification of an impurity can lead to an understanding of its physicochemical properties. This, in turn, can provide information for process modifications and the subsequent elimination of the impurity from future drug lots. Secondly, an impurity's identity can provide assurance to the FDA and individuals conducting safety and clinical trials as to the possible toxicity, carcinogenicity, or teratogenicity of the impurity. Thirdly, any effect that impurities have on the drug, such as proteolytic degradation, can be examined after their identification. Finally, identification influences the development of specific assays to monitor an impurity for lot-to-lot drug consistency.

Our objective in this chapter is to review the chromatographic procedures used by bioanalysts to assay recombinant protein therapeutics for host-cell protein impurities. Generally, these procedures involve either electrophoresis or high performance liquid chromatography (HPLC). We will also review current technology used in conjunction with these procedures to specifically identify protein impurities. A short discussion is also included that examines the applicability of capillary electrophoresis to the analysis of host cell-protein impurities.

Gel Electrophoresis

Gel electrophoresis is an indispensable form of separation and analysis for the bioanalyst. The application of an electric current induces proteins to migrate according to their overall charge through a semisolid medium generally consisting of crossed-linked polyacrylamide. The polyacrylamide gel acts as a sieve that differentially impedes proteins based on their molecular size and shape. The electrical current provides the driving force similar to more traditional mobile phases while the gel's sieving effect acts as a retarding force. As in all forms of chromatography, separation is the result of these two forces interacting on individual analytes.

Numerous forms of gel electrophoresis exist for the analysis of proteins (*3*), but polyacrylamide gel electrophoresis in the presence of sodium dodecyl sulfate (SDS-PAGE), popularized by Laemmli (*4*), predominates. SDS is an anionic detergent that denatures and binds proteins in a ratio of 1.4 g of SDS per gram of protein (*5*). SDS denaturation produces a similar shape and a constant charge density for all protein molecules. These changes make molecular size the sole basis of protein separation by this method. SDS-PAGE is often the first choice for analyzing recombinant proteins for purity (*2*).

Electrophoretically separated proteins are detected by staining. The two most common stains are Coomassie Brilliant Blue R-250 and silver. The mechanism of Coomassie Blue staining is not fully understood. It is thought that the dye binds primarily to basic and aromatic residues in proteins (*6*). The major advantage of Coomassie Blue is the almost universal quantitative binding of the dye to proteins. In a study of Coomassie Blue binding to twenty-six proteins in solution, staining variability was less than two-fold (*6*).

Protein impurities are detected and quantitated electrophoretically when known amounts of the drug product are subjected to SDS-PAGE followed by Coomassie Blue staining. Although slight variations are common, generally 2 - 14 μg of the drug product are within the linear range of the assay. We have found some impurities are detected at levels as low as 60 ng. This yields a detection limit of 0.6% for a 10 μg sample (6000 ppm). Since the linear range of the assay must not be exceeded, analyzing more than 10 μg of protein only increases the detection limit slightly. Once stained, protein impurities are estimated by visually comparing their staining intensity to that of known amounts of standard proteins electrophoresed through the same gel. Quantitation of protein impurities is more accurately determined by scanning the stained gel with a laser densitometer to produce an electropherogram. The area percent of each peak in the electropherogram is determined by methods identical to those of HPLC chromatograms. Figure 1 demonstrates this process. Samples of recombinant acidic fibroblast growth factor (aFGF) were spiked with known amounts of a *E. coli* protein impurity (S3 ribosomal protein) and electrophoresed through a 15% SDS-polyacrylamide gel. After electrophoresis, the gel was stained with Coomassie Blue and scanned with a laser densitometer. The detection limit for this particular protein impurity was 200 ng or 0.7%.

The mechanism of silver staining proteins is not entirely understood (*7*). In general, silver staining is more sensitive than Coomassie Blue staining. Protein levels as low as 0.2 ng have been detected (*8*). If comparable sensitivity was universal among proteins then a protein impurity as low as 0.002% (20 ppm) could be detected by assaying 10 μg of total protein. Unfortunately, silver staining is highly variable and thus does not facilitate accurate quantitation (*8,9*). In our lab, we have analyzed one protein that is marginally detected at 125 ng by silver and not detected at all at 50 ng. If a protein impurity showed this same sensitivity then it could not be detected at a level of 1.25% ($>10^4$ ppm). This level of detection is equivalent to Coomassie Blue staining. When the two staining procedures are used simultaneously for model proteins the sensitivity is reportedly

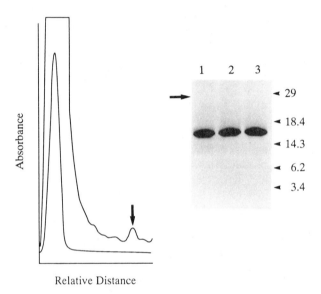

Figure 1. Electropherograms and SDS-PAGE of recombinant aFGF spiked with an *E. coli* protein impurity. Left: Two electropherograms of an identical sample of 30 μg of aFGF spiked with 500 ng of *E. coli* S3. The peak representing the S3 impurity was observed when the absorbance scale was expanded 30-fold. The 30 μg of aFGF was within the linear range of this assay. Right: The Coomassie Blue-stained gel. Each lane contained 30 μg aFGF spiked with: 1) 500 ng of S3, 2) 200 ng of S3, and 3) no S3. The location of the S3 impurity is indicated by the arrows. The migration of molecular weight markers is indicated in kilodaltons.

increased two to eight-fold compared to silver alone (*7*). This variable enhancement might also extend to protein impurities. The high sensitivity of silver staining is itself problematic. To detect a protein impurity at 20 ppm requires analyzing at least 10 μg of the recombinant protein. Large protein loads obscure portions of the gel and possibly conceal impurities migrating close to the therapeutic. In addition to silver and Coomassie Blue detection, other stains exist, but their use is less widespread (*3*).

Coomassie Blue and silver staining of protein gels are the standard electrophoretic methods used to detect and often quantitate protein impurities in recombinant therapeutics. Both these techniques suffer from an identical drawback, that is they lack specificity. An impurity detected by either approach is uncharacterized. The impurity can be a host-cell protein or another protein impurity such as an aggregate or degradate of the drug product. Host-cell impurities are specifically identified by employing an electrophoretic immunoassay. The most commonly used method is western blotting first popularized by Burnette a decade ago (*10*).

The basic methodology of western blot analysis starts with SDS-PAGE. After electrophoresis, proteins are electrotransferred (blotted) from the gel to an immobilizing matrix (*11*). The two most common matrices are nitrocellulose paper and polyvinylidene fluoride (PVDF). Proteins are irreversibly bound to the matrix and then detected immunochemically. In direct analysis, immobilized proteins are probed with either radiolabeled antibodies, or antibodies conjugated to a fluorophore or an enzyme. If the protein is recognized by the antibody, the antibody binds the protein and is detected according to the appropriate signal (*12*). Often this approach is not practical and indirect techniques are required. Here, the immobilized proteins are first probed with specific unlabeled antibodies. The bound antibodies are then detected by a secondary probe consisting of either labeled antibodies or labeled Protein A.

Western blot analysis is the most specific and sensitive assay available to the bioanalyst. As little as 30 pg of protein have been detected (*13*). With a total protein load of 10 μg this yields a sensitivity of 3 ppm. This level of sensitivity is adequate by most criteria. Unfortunately, this sensitivity does not prevail for all host-cell protein impurities. The antibodies used in these westerns are generally raised against a lysate derived from host cells that do not produce the recombinant therapeutic. A portion of the proteins in these lysates are only weakly reactive to the antiserum. Some proteins may not elicit an antibody response at all because their concentration is too low or they are poorly immunogenic. Hence, this technique may not be able to detect all protein impurities. Gooding and Bristow used western blot analysis to detect trace protein impurities in preparations of recombinant human growth hormone (*14*). They found that highly antigenic *E. coli* proteins were detected at a greater sensitivity in western blots than by silver staining, but the latter technique detected more protein impurities. Western blots are also poorly quantitative, but using antiserum that specifically recognizes a particular protein impurity provides limited quantitation. Recent work has led to the development of non-chromatographic immunoassays that quantitate protein impurities using host-cell antiserum (8,15). Although lacking quantitation, westerns are valuable tools for showing lot-to-lot drug consistency.

Western blot analysis has additional shortcomings. Proteins are electrotransferred from polyacrylamide gels with variable efficiency. Transfer to the immobilizing matrix is inversely related to the protein's size; large molecules transfer less efficiently than small molecules. The inclusion of 0.1% SDS in the transfer buffer or limited proteolysis within gels enhance the transfer of large proteins (12). Small proteins bind less effectively to the immobilizing matrix than do large proteins. This drawback is reportedly overcome by using chemical crosslinkers to covalently bind the protein to the transfer matrix (16). These problems, as well as the general non-quantitative nature of westerns, demonstrate that this technique by itself cannot provide a complete impurity profile. This underscores the need to use both silver and Coomassie Blue staining in conjunction with western blot analysis (1,17).

We have taken this multi-pronged approach for analyzing host-cell protein impurities in our laboratory. Recombinant aFGF produced at Merck is >99% pure when analyzed by SDS-PAGE. Protein impurities were undetected by Coomassie Blue staining (Figure 1, lane 3), but occasionally in some early production lots a low level impurity of ~25 kD was detected by silver staining. When antiserum raised against an E. coli cell lysate was used in a western blot of aFGF the source of this impurity proved to be the host cells (Figure 2). Through a combination of E. coli western blot analysis and HPLC, samples enriched in the ~25 kD impurity were prepared. The N-terminal sequence of the impurity was determined in an automated protein sequencer. The result was compared to the National Biomedical Research Foundation (NBRF) protein database and was found to match that of the S3 ribosomal protein of E. coli. We confirmed the identity of the impurity as the S3 protein using S3 monoclonal antibodies in a western blot of aFGF. The level of the S3 protein was estimated by the same technique using the enriched S3 sample as a standard. The amount of S3 in a number of aFGF preparations was shown to be less than .01% (<100 ppm) by this method. Western blot analysis was also performed on a recombinant antibody produced in mammalian cells (Figure 2). Coomassie Blue staining of SDS-polyacrylamide gels showed little evidence of host-cell protein impurities, but western blot analysis using antiserum raised against the mammalian host-cell lysate showed the presence of at least one host-cell protein impurity of ~97 kD (Figure 2). The identity and level of this impurity has not yet been determined.

High Performance Liquid Chromatography

HPLC is recommended and frequently used for purity analysis of recombinant proteins (1). Specifically, levels of host-cell proteins and cell culture or fermentation medium proteins can be monitored. HPLC is used to quantitate protein impurities as low as 0.1% (18). The advent of columns containing small particle (3-5 μm), wide pore (300 Å) C_1 to C_{18} bonded silica has increased the chromatographic efficiency of protein separations making reversed-phase HPLC (RP-HPLC) the primary method for impurity profiling. Hydrophobic interaction

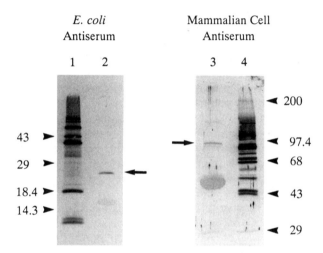

Figure 2.   Western blot analysis of recombinant proteins using antisera recognizing host-cell proteins. Left panel: Recombinant aFGF (lane 2) and a 100-fold dilution of an *E. coli* cell lysate (lane 1) were electrophoresed through a 12% SDS-polyacrylamide gel, electroblotted to nitrocellulose, and then probed with *E. coli* antiserum.   Right panel: A recombinant antibody (lane 3) and a 100-fold dilution of a mammalian host-cell lysate (lane 4) were similarly analyzed using a 7% SDS-polyacrylamide gel and mammalian cell lysate antiserum.   Host-cell protein impurities are indicated by the arrows. The migration of molecular weight markers for each blot are shown in kilodaltons.

(HIC), ion exchange (IEX) and size exclusion (SEC) chromatographies are also employed but to a lesser extent (8).

In RP-HPLC distinct differences in the hydrophobicity of each protein component and their interactions with the alkyl side chains of the stationary phase determine separation. These differences in hydrophobicity must occur in the "contact region" of the proteins to affect separation. Thus, this adsorption mechanism is highly dependent on conformation. In the native conformation of most proteins, hydrophobic sequences are sequestered internally while exposed protein sequences tend to be more hydrophilic.

The most widely used mobile phase system employs a water/acetonitrile gradient containing 0.1% trifluoroacetic acid. The starting conditions and the gradient itself vary with each separation. Other mobile phase systems employ phosphoric acid, acetate, or phosphate buffers as the aqueous phase and methanol, 1-propanol or 2-propanol as the organic phase. Each mobile phase system leads to differential protein denaturation and the exposure of hydrophobic sequences. The extent of denaturation is dependent on the pH and the amount and the type of organic phase. These changes influence the protein's contact with the stationary phase and thus separation. The denatured or partially denatured proteins adsorb to the hydrophobic surface of the stationary phase where they remain until desorbed by a "critical" concentration of organic phase (19). During desorption the protein can be further denatured. Small changes in pH and the percent and type of organic phase influence a protein's retention. Proteins can display a number of sites capable of interacting with the stationary phase. These multiple interactions can lead to several distinctly resolved forms of a protein. The degree of column end-capping also influences protein separation since the silanol groups can function as cationic sites leading to a mixed mode separation. The recent introduction of polymeric supports eliminates this problem and allows the use of high pH mobile phases. Care must be taken in developing methods that meet the desired requirements.

HIC, an extension of reversed-phase chromatography, also separates proteins by differential hydrophobic interaction with the stationary phase. Unlike RP-HPLC, protein structure is preserved by using aqueous salt gradients as mobile phases. Samples are injected onto the column equilibrated in high salt. These conditions enhance the protein's hydrophobic interaction with the stationary phase by removing water molecules from the protein's surface. A descending salt gradient leads to rehydration and selective release from the stationary phase maintaining the protein's native state.

IEX separates proteins on the basis of charge. Samples acquire a net charge from the initial mobile phase pH and are applied to an oppositely charged stationary phase. A reversible electrostatic interaction binds the proteins to the stationary phase. Either a gradient of increasing ionic strength or a pH gradient elute the proteins from the stationary phase. This technique allows for a vast range of variables that influence selectivity.

Separation by SEC is based on a protein's hydrodynamic volume and its interaction with the porous stationary phase. Large proteins are excluded from the pores of the particles and are the first to elute. Other proteins enter the particles based on their size. Small proteins access more of the particle's interior and therefore are retained longer and eluted later. Isocratic aqueous mobile phases are generally used. Proteins can be chromatographed in either the denatured or native state.

Of the four chromatographic techniques described, reversed-phase HPLC provides the greatest selectivity. Methodology development and operation are also easier with RP-HPLC. Although IEX also provides great potential selectivity, it can be difficult to develop. SEC and HIC are limited by their lack of high resolution. Yet, these chromatographic modes aid in the physicochemical characterization of protein impurities possibly leading to process modifications. The best method for host-cell protein impurity analysis must be determined on a case-by-case basis.

The most common method of detection for these chromatographic techniques is UV absorbance. Caution is needed in selecting mobile phase components because the UV absorbance of the system must be minimized. Greatest sensitivity is obtained between 200 nm and 220 nm due to the protein's peptide bonds. The absorbance is proportional to the number of peptide bonds present so the quantity of an impurity is determined using a standard protein. Unfortunately, non-protein impurities can also absorb in the UV range so further analysis is required to identify an impurity as a protein. Detection of the amino acid chromophores phenylalanine, tyrosine, and tryptophan at 257 nm, 275 nm, and 280 nm, respectively, helps establish an impurity as a protein. Diode array detection allows monitoring of the full UV spectrum for this purpose. However, these detectors are typically ten-fold less sensitive than fixed wavelength detectors so their use in host-cell protein impurity analysis is suspect. Fluorescence detection is also used with high sensitivity to monitor the fluorescence of phenylalanine, tyrosine, and tryptophan. Again, this helps identify an impurity as a protein, but the unequal distribution of these residues in proteins and the dependence of their fluorescence on protein conformation limits their use in quantitative analysis.

Figure 3 demonstrates the practical application of RP-HPLC to protein impurity analysis. A purified recombinant therapeutic was spiked with known amounts of a protein impurity, subjected to RP-HPLC and monitored for absorbance at 218 nm. The mass detection limit of the protein impurity under these conditions was ~0.02 μg. This, coupled with an upper linear limit of at least 49 μg for the therapeutic, yields a sensitivity of ~0.04% (~400 ppm). This sensitivity is not only 10-fold greater than that of Coomassie Blue-stained gels, but it also offers the advantage of automation.

All the HPLC systems and detectors described are incapable of specifically identifying host-cell protein impurities. Establishing an impurity as a protein is only the first step. The protein impurity can be an altered form of the drug product and not a host-cell protein. Recently though, new technology is emerging that couples the automation and speed of HPLC systems with the specificity of western blot analysis. This technology is called high performance immunoaffinity chromatography (HPIAC) (20). In this mode of HPLC, specific antibodies are immobilized on the stationary phase. For analyzing host-cell protein impurities, the immunoadsorbent would be antiserum raised against a host-cell lysate. Similar to western blots, this approach is restricted by the limitation of the

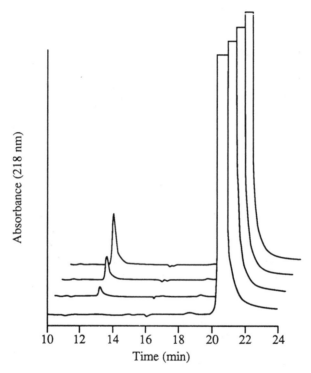

Figure 3. RP-HPLC of a recombinant therapeutic spiked with a protein impurity. Samples of a recombinant therapeutic (9 μg) were spiked with different amounts of a known protein impurity. From top to bottom, the chromatograms were spiked with 0.2, 0.09, 0.04, and 0.0 μg of the protein impurity. Column: VYDAC C$_4$ (.46 x 15 cm), 5 μm, 300 Å. Linear gradient: 32% - 68% acetonitrile in 0.1% trifluoroacetic acid over 40 min at 1.0 ml/min. Detection: 218 nm.

antiserum to contain antibodies recognizing all potential impurities. HPIAC can be used as the initial step in multi-dimensional chromatographic analysis. Host-cell protein impurities captured and concentrated by immunoadsorption are then transferred to a reversed-phase system by column switching for further analysis (*21*).

Identification of Host-Cell Protein Impurities

Once a protein impurity is detected it can be identified. The most direct method of identification begins with N-terminal sequence analysis. The gas-liquid solid-phase sequencer routinely determines the N-terminal sequences of proteins at the 10 to 100 picomole level. Samples are first immobilized on a glass fiber disc and then subjected to repetitive Edman degradations. The released amino acids are identified by reversed-phase HPLC. The entire process is automated and proceeds with no operator input (*22*).

The level of the protein impurity influences the steps necessary to identify it. A high level impurity can be sequenced directly from a PVDF membrane. The impurity is first electrophoretically resolved from the drug product by SDS-PAGE and then electrotransferred to the PVDF membrane (*23*). The impurity is visualized by staining with Coomassie Blue R-250, Amido Black or Ponceau S. The stained impurity is excised and placed directly into the micro-sequencer for N-terminal sequencing. The identification of ten to twenty N-terminal amino acids is generally sufficient. The sequence obtained is then compared to a database, such as the NBRF database, to determine the identity of the protein impurity. If the impurity is present in trace amounts (<10 ppm) it is necessary to prepare a chromatographically enriched sample from process chromatographic side-cuts prior to sequence analysis. We employed this latter approach to identify the bacterial protein impurity in preparations of recombinant aFGF (Figure 2).

If an automated sequencer is not available or the impurity's sequence is not in the database, two-dimensional gel electrophoresis can prove helpful. A sample containing the impurity is electrophoretically separated in the first dimension based on its isoelectric point (isoelectric focusing). The second dimension is SDS-PAGE and separates the proteins according to their size (*24*). After staining, the relative position of the protein impurity is referenced by coordinates to a two-dimensional gel database of *E. coli* proteins (*25*). An isoelectric point and molecular size similar to that of a database protein suggests the two proteins are the same. If the protein in the gel database is known then specific antibodies can be developed to identify and quantitate the protein impurity in the drug product. Unfortunately, only ~10% of *E. coli* proteins are currently identified in the database.

Capillary Electrophoresis

Although many forms of HPLC are available for the analysis of recombinant protein therapeutics, often electrophoresis is the final test for evaluating purity (*26*). The limitations in sensitivity, quantitation, and reproducibility, plus the intense labor required for gel electrophoresis, have created the need for better analytical methods. This need can possibly be met by the emerging technique of capillary electrophoresis (CE) (27).

CE achieves high resolution separations based on the differential migration of analytes in an electric field. CE analysis of proteins has often been carried out in open tubular capillaries, a mode pioneered by Jorgenson and Lukacs (28). In this mode, a buffered-filled capillary links two reservoirs. When samples are injected into the capillary they migrate through the tube when an electric current is applied. A detector is located at one end of the tube for monitoring the analytes. The mobility of each analyte is dependent on size, shape, and charge (27). If the size and shape of impurities are equivalent then separation is dictated by charge differences.

The application of CE to protein therapeutic purity analysis has been limited. Hurni and Miller used CE to monitor host-cell protein impurities during the purification of a recombinant vaccine (29). These authors developed a host-cell protein impurity profile for each step in the purification process to demonstrate lot-to-lot consistency. CE has primarily focused on detecting impurities closely related to the drug product. Such impurities include deamidated proteins where a charge difference results due to the introduction of a carboxyl group. Sulfoxides of the amino acid methionine have also been resolved (30). Generally, this analysis proceeds after the protein therapeutic is specifically fragmented by an enzyme. Impurities in both recombinant human growth hormone and biosynthetic human insulin have been demonstrated by this method (18,31).

CE analysis of host-cell protein impurities has not been extensively developed, but applications will appear due to the technique's advantages. These include automation, high resolution, and sensitive detection and quantitation. Similar to HPLC, UV absorbance is the most common method of detection for proteins in CE. Wavelengths as low as 190-200 nm provide adequate sensitivity for most analyses (32). Nielsen and Rickard were able to observe a 0.4% (4000 ppm) impurity with a 25 ng sample load of human recombinant growth hormone (18). Proteins are also detected fluorescently in CE. As noted earlier for HPLC, this provides increased sensitivity, but protein fluorescence is variable. Further improvements in sensitivity are potentially obtained when protein samples are derivatized with a fluorophore (27). Nickerson and Jorgenson used this approach and detected 44 attomoles of a protein (33). If this same sensitivity were to hold for the analysis of host-cell proteins then impurities as low as 30 ppm could be detected. For analyzing trace protein impurities, this level of detection is approximately equivalent to the most sensitive western blots. Coupling protein derivatization with laser-induced fluorescence can potentially increase the sensitivity of CE beyond that of western blots (18,30). When samples fluoresce poorly and derivatization is ineffective then indirect fluorescence detection is an alternative. In this method, a fluorophore is incorporated in the CE buffers and analytes are detected when the fluorophore is displaced (27). The mass detection limit for peptides in indirect fluorescence CE is almost 200-fold less than UV absorbance detection (34). Indirect fluorescence has the potential to be a universal, highly sensitive method for detecting protein impurities because it depends on neither intrinsic protein fluorescence nor derivatization.

Summary and Future Considerations

Clearly, no single technique has the capability to detect, identify, and quantitate all the host-cell protein impurities in any recombinant pharmaceutical product. The most prudent approach involves a combination of existing methodologies. The continual pursuit of therapeutics from recombinant DNA technology plus the delivery of purer products demands technological improvements in the areas of sensitivity, specificity, and automation. One approach to reaching these objectives is to integrate present methodologies. CE and HPIAC are two examples that are moving towards this goal, another is HPLC with on-line mass spectroscopic detection. These techniques and others will be welcome additions to the bioanalysts growing armamentarium.

Acknowledgments

We are grateful to Dr. Larry Kahan of the University of Wisconsin for supplying us with monoclonal antibodies recognizing the S3 protein. We also wish to thank Mr. Peter DePhillips for supplying us with the S3 protein and Mr. Timothy Smith and Mr. Anthony Paiva for analysis of polyacrylamide gels.

Literature Cited

1. *Points to Consider in the Production and Testing of New Drugs and Biologicals Produced by Recombinant DNA Technology;* Office of Biologics Research and Review, Center for Drugs and Biologics, FDA, 1985.
2. Bogdansky, F. M. *Pharm. Technol.* **1987**, *11*, 72-74.
3. Andrews, A. T. *Electrophoresis: Theory, Techniques, and Biochemical and Clinical Applications;* Oxford University Press: NY, NY, 1986; Second Edition.
4. Laemmli, U. K. *Nature* **1970**, *227*, 680-685.
5. Reynolds, J. A.; Tanford, C. *J. Biol. Chem.* **1970**, *245*, 5161-5165.
6. Davis, E. M. *Amer. Biotechnol. Labor.* **1988**, *July*, 28-36.
7. De Moreno, M. R.; Smith, J. F.; Smith, R. V. *Anal. Biochem.* **1985**, *151*, 466-470.
8. Garnick, R. L.; Solli, N. J.; Papa, P. A. *Anal. Chem.* **1988**, *60*, 2546-2557.
9. Oakley, B. R.; Kirsch, D. R.; Morris, N. R. *Anal. Biochem.* **1980**, *105*, 361-363.
10. Burnette, W. H. *Anal. Biochem.* **1981**, *112*, 195-203.
11. Towbin, H.; Staehelin, T.; Gordon, J. *Proc. Natl. Acad. Sci. USA* **1979**, *76*, 4350-4354.
12. *CRC Handbook of Immunoblotting of Proteins;* Bjerrum, O. J.; Heegaard, N. H. H., Eds.; CRC Press, Inc.: Boca Raton, FL, 1988; Vol. 1.
13. Blake, M. S.; Johnston, K. H.; Russel-Jones, G. J.; Gotschlich, E. C. *Anal. Biochem.* **1984**, *136*, 175-179.
14. Gooding R. P.; Bristow, A. F. *J. Pharm. Pharmacol.* **1985**, *37*, 781-786.

15. Anicetti, V. In *Analytical Biotechnology: Capillary Electrophoresis and Chromatography;* Horváth, C; Nikelly, J. G., Eds.; ACS Symposium Series 434; American Chemical Society: Washington, DC, 1990; 127-140.
16. Kakita, K.; O'Connell, K.; Permutt, M.A. *Diabetes* **1982,** *31,* 648-652.
17. Anicetti, V. R.; Keyt, B. A.; Hancock, W. S. *Trends Biotechnol.* **1989,** *7,* 342-349.
18. Nielsen, R. G.; Rickard, E. C. In *Analytical Biotechnology: Capillary Electrophoresis and Chromatography;* Horváth, C; Nikelly, J. G., Eds.; ACS Symposium Series 434; American Chemical Society: Washington, DC, 1990; 36-49.
19. Geng, K.; Regnier, F. E. *J. Chromatogr.* **1984,** *296,* 15-30.
20. Phillips, T. M. In *Advances in Chromatography;* Giddings, J. C.; Grushka, E.; Brown, P. R. Eds.; Marcel Dekker, Inc.: NY, NY, 1989, Vol. 29; 133-173.
21. Regnier, F. E.; Xu, C.; de Frutos, M.; Birck-Wilson, E.; Dorval, B.; Afeyan, N. *Proceedings of the Eleventh International Symposium on HPLC of Proteins, Peptides, and Polynucleotides;* Washington, DC, October 20-23, 1991.
22. Hewick, R. M.; Hunkapiller, M. W.; Hood, L. E.; Dreyer, W. J. *J. Biol. Chem.* **1981,** *256,* 7990-7997.
23. Matsudaira, P. *J. Biol. Chem.* **1987,** *262,* 10035-10038.
24. Garrels, J. I. In *Methods in Enzymology;* Wu, R.; Grossman, L.; Moldave, K., Eds.; Academic Press: NY, NY, 1983, Vol. 100; 411-423.
25. Phillips, T. A.; Vaughn, V.; Bloch, P. L.; Neidhardt, F. C. In *Escherichia coli and Salmonella typhimurium: Cellular and Molecular Biology;* Neidhardt, F. C., Ed.; American Society for Microbiology: Washington, DC, 1987, Vol. 2; 919-966.
26. Guzman, N.A.; Hernandez, L.; Terabe, S. In *Analytical Biotechnology: Capillary Electrophoresis and Chromatography;* Horváth, C; Nikelly, J. G., Eds.; ACS Symposium Series 434; American Chemical Society: Washington, DC, 1990; 1-35.
27. Gordon, M. J.; Huang, X.; Pentoney, Jr., S. L.; Zare, R. N. *Science* **1988,** *242,* 224-228.
28. Jorgenson, J. W.; Lukacs, K. D. *Anal. Chem.* **1981,** *53,* 1298-1302.
29. Hurni, W. M.; Miller, W. J. *J. Chromatogr.* **1991,** *559,* 337-343.
30. Karger, B. L.; Cohen, A. S.; Guttman, A. *J. Chromatogr.* **1989,** *492,* 585-614.
31. Nielsen, R. G.; Sittampalam, G. S.; Rickard, E. C. *Anal. Biochem.* **1989,** *177,* 20-26.
32. Novotny, M. V.; Cobb, K. A.; Liu, J. *Electrophoresis* **1990,** *11,* 735-749.
33. Nickerson, B.; Jorgenson, J. W. *J. Chromatogr.* **1989,** *480,* 157-168.
34. Hogan, B. L.; Yeung, E. S. *J. Chromatogr. Sci.* **1990,** *28,* 15-18.

RECEIVED April 7, 1992

Chapter 11

Binding Proteins in Development of On-Line Postcolumn Reaction Detection Systems for Liquid Chromatography

Minas S. Barbarakis and Leonidas G. Bachas[1]

Department of Chemistry, University of Kentucky, Lexington, KY 40506–0055

The use of binding assays for analyte detection following HPLC separation provides a powerful tool for the determination of metabolites in biological samples. Such an approach, combines efficiently the superior selectivity of HPLC with the high detection capability of binding assays. In addition, it allows for multiple analyte determinations in the same sample. Recent advances in this area regarding the use of the binding protein avidin in the development of on-line postcolumn reaction detection systems demonstrate the merits of the technique.

Biochemical interactions, such as those of binding proteins with ligands, enzymes with substrates, antibodies with antigens, receptors with hormones, etc., are characterized by high degree of specificity. In all these cases, specificity is achieved by formation of noncovalent complexes where at least one of the reacting species is macromolecular. Such biospecific interactions form the basis of a range of chromatographic techniques (e.g., affinity chromatography) (1, 2) and binding assays (e.g., immunoassays, protein binding assays, etc.) (3).

High performance liquid chromatography (HPLC) plays a predominant role in the analysis of a wide variety of samples and provides a highly selective, rapid, reproducible, and automatable technique (4). Specifically, HPLC is considered to be a powerful tool in the pharmaceutical industry which includes the drug discovery process (e.g., isolation of natural products), course of drug development (e.g., purity determinations and assays), and process validation and control (5). Although HPLC provides high selectivity, the sensitivity of the associated detection systems does not always allow for trace-level determinations, which are sometimes required in biomedical analysis. Specifically, the conventional HPLC detectors have limited detection capabilities when analyzing compounds that do not contain strong chromophores, fluorophores, or electrochemically active groups. Although the use of more sensitive detectors based on mass spectrometry (6) or laser technology (7) is possible, much of the instrumentation reported to date is expensive and/or complex.

---

[1]Corresponding author

On the other hand, binding assays have experienced a wide applicability in the determination of a variety of important physiological substances such as drugs, mainly because of their sensitivity and detection limits, which are typically in the nanogram-to-picogram range. Although binding assays are considered to be highly sensitive techniques, their specificity is limited when dealing with samples that contain biomolecules that are structurally similar to the analyte (e.g., metabolites) and which can interfere (i.e., cross-react) with the biological binder used in the assay. For instance, the cross-reactivity of antisera employed in radioimmunoassays that are used for morphine determinations in plasma decreases the specificity of these analytical methods, because in patients receiving chronic doses of morphine, the levels of the drug metabolites are many times greater than that of the parent drug (8). Consequently, it is necessary to include purification steps in the analysis of samples that contain both the analyte and its metabolites.

From the argument above it appears that a coupled HPLC-binding assay procedure should be ideal for the determination of metabolites in a biological sample. In such a case, the binding assay would detect and quantify with high sensitivity the HPLC-separated components in biological extracts, while at the same time minimizing the drawbacks associated with insufficient detectability in HPLC and low specificity in binding assays.

It should be noted that coupled HPLC-binding assay procedures exhibit an additional advantage. Specifically, binding assays alone can not determine multiple analytes (e.g., drug and its metabolites) in a single run (usually only one analyte is determined at a time), whereas HPLC can. Indeed, numerous HPLC-based analyses of drugs and their metabolites have been reported, such as those for opiates (9), digitoxin (10), and theophylline (11). The coupled HPLC-binding assay techniques reported here allow the determination of multiple analytes in the same sample, thus minimizing sample consumption.

Coupling HPLC and Binding Assays

The combination of HPLC with binding assays can be carried out in an off-line or an on-line mode. Although both modes of performance are as selective and sensitive, the total time of analysis for the off-line mode is longer. Therefore, it is surprising that the majority of the reported HPLC-binding assay procedures are employed in an off-line mode (12-20). This chapter will focus on the use of binding proteins for the development of on-line postcolumn reaction detection systems for liquid chromatography. In addition, emphasis will be given to the time involved, cost and convenience of the reported techniques.

On-line postcolumn binding assay methods can be carried out in various formats depending on the type of binding assays involved (3). In general, binding assays employing biochemical interactions for the determination of analytes involve at least one reaction and a detection step. These assays can be classified according to (a) the type of binder employed (e.g., antibody, binding protein, etc.); (b) the use or not of a physical separation of the "bound" and "free" biochemical reacting species that classifies these assays as heterogeneous and homogeneous, respectively; (c) the detection method used (e.g., fluorescence, radioactivity, etc.); and (d) whether the analyte competes with a labeled component for the binder active site giving rise to competitive (indirect) or noncompetitive (direct) assay.

Three different postcolumn detection schemes were developed recently in our laboratory for the on-line detection of model analytes (i.e., biotin and biocytin) separated by HPLC. In all three cases the detection scheme was based on the selective and strong interaction between the binding protein avidin and its specific ligand biotin (the binding constant is on the order of 10^{15} M^{-1}). Two of the postcolumn reaction detection schemes involved competitive binding assays with photometric and fluorescence detection, and the third one used the avidin-biotin system as part of a direct fluorescence-based assay. Experimentally, the effluent from the HPLC column (in our case, a C_{18} reverse phase column) is merged with an appropriate reagent phase, permitted to react in an on-line mode, and detected in a proper manner (Figure 1). A description and the merits of the three postcolumn detection systems are given below.

It should be noted that the principle of this approach is different from that of affinity chromatography. Therefore, the regeneration step typically required in affinity chromatography (21, 22) is avoided and the sample throughput, which is of considerable significance in biochemical analyses, is greatly increased. Further, the approach described in this proposal should be distinguished from the work of Heineman, Halsall and co-workers (23). These investigators have used HPLC with electrochemical detection in order to observe the formation of the enzymatic product of an immunoassay reaction (e.g., NADH) in the presence of electroactive interferents (e.g., uric acid when analyzing urine). In that respect, their system cannot be considered as a postcolumn reaction and the HPLC is not used to separate the analytes (i.e., the problems of cross-reactivity mentioned before still exist in their system).

Avidin-Biotin Complex in the Development of On-Line HPLC-Binding Assay Detection Systems

System I. This reported postcolumn reaction detection scheme (24) took advantage of the following facts: (a) the ability of avidin to bind the dye 2-[4'-hydroxyphenyl-azo]benzoic acid (HABA) with a concomitant decrease in the absorbance signal of the free dye at 345 nm and the appearance of a new band at 500 nm (25), and (b) the ability of biotin and biocytin to displace the dye from the avidin-HABA complex because the binding constant of the avidin-biotin/biocytin complex (i.e., ca. 10^{15} M^{-1} for biotin (26)) is significantly higher than that of the avidin-HABA complex (i.e., ca. 10^5 M^{-1} (25)). This displacement results in an increase in the absorbance signal at 345 nm and a decrease in absorbance at 500 nm. The two competing equilibria in this system are shown below:

$$\text{biotin (or analogue)} + \text{avidin} \rightleftharpoons \text{avidin-biotin (or avidin-analogue)}$$
$$\text{HABA} + \text{avidin} \rightleftharpoons \text{avidin-HABA}$$
$$\lambda_{max} = 345 \text{ nm} \qquad\qquad \lambda_{max} = 500 \text{ nm}$$

In the instrumental setup described in Figure 1, the effluent stream from the HPLC column is mixed on-line with a second stream containing the avidin-HABA complex, and the analyte concentration is determined by monitoring the absorbance of the displaced free dye at 345 nm with a UV-vis detector. The reported postcolumn reaction detection scheme improved both the selectivity and sensitivity of the biotin and biocytin determinations (24). Figure 2 shows the chromatograms of 20 µL of a mixture of biotin and biocytin with direct UV detection at 220 nm (Figure 2A) and using the avidin-HABA complex (Figure 2B) under the same elution conditions. A comparison of the two chromatograms demonstrates that the competitive binding principle can indeed be employed for the development of on-line postcolumn reactors for analyte detection.

System II. This postcolumn reaction detection scheme is in principle similar to the one above. The difference lies in the use of the fluorescent probe 2-anilino-naphthalene-6-sulfonic acid (2,6-ANS) instead of HABA. Upon binding to avidin, the fluorescence of 2,6-ANS exhibits a blue shift with a large increase in quantum yield. In the presence of biotin derivatives, the probe is displaced from the

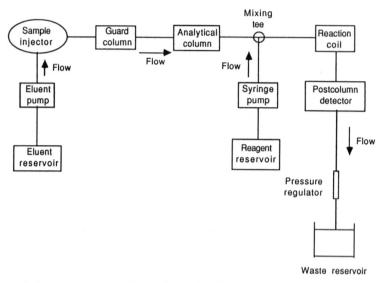

Figure 1. Instrumental setup for on-line postcolumn reaction detection in HPLC.

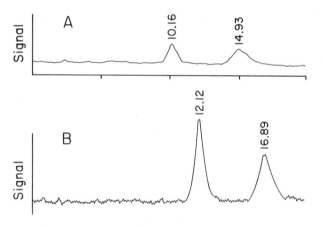

Figure 2. Chromatograms of 20 μL of a mixture of biotin and biocytin (2.0 x 10^{-4} M) (A) with direct UV detection at 220 nm, and (B) with a postcolumn reaction detection scheme using the avidin-HABA complex. Both chromatograms are on the same scale. (Reproduced from ref. 24. Copyright 1990 American Chemical Society.)

avidin-2,6-ANS complex resulting in a decrease of the fluorescence at 438 nm. The two competing equilibria in this system are:

biotin (or analogue) + avidin $\rightleftharpoons$ avidin-biotin (or avidin-analogue)

2,6-ANS + avidin $\rightleftharpoons$ avidin-2,6-ANS

$$\lambda_{ex} = 328 \text{ nm} \qquad\qquad \lambda_{ex} = 328 \text{ nm}$$

$$\lambda_{em} = 462 \text{ nm} \qquad\qquad \lambda_{em} = 438 \text{ nm}$$

The use of fluorescence as the detection mode provided higher sensitivity (compared to the spectrophotometric detection in System I) for the HPLC determinations of biotin and biocytin (27). Additional advantages of this system include the high stability of the fluorescent probe and its large Stokes shift (28).

The selectivity of the method was examined by analyzing solutions containing biotin, biocytin and other interferents that have similar retention times as biotin and biocytin, and absorb at 220 nm. As it is shown in Figure 3, the direct detection at 220 nm suffered from severe interference that precluded the correct determination of either biotin or biocytin (Figure 3A). However, by using the postcolumn reaction system the interferents did not influence the detector response (Figure 3B). In addition, the reported postcolumn reaction detection system was validated by determining biotin in commercial vitamin preparations and in extracts of a horse-feed supplement (27). The results of the analyses shown in Table I exhibit an accuracy of 3-6% for the biotin determinations.

Table I. Analyses of real samples (adapted from ref. 27)

Sample	Source	Units	Amount of biotin Found [a]	Claimed
Biotin	Wm. T. Thompson	µg per tablet	161	150
Biotin	Perrigo	µg per tablet	290	300
Horse-feed supplement	Nickers International	mg Kg^{-1}	230	231

[a] Average of two determinations.

As it is shown in Figure 4, the direct UV detection at 220 nm (Figure 4A) does not allow for correct determination of biotin in the extract from the horse-feed supplement due to interferences from other matrix components. However, the postcolumn reaction with fluorimetric detection at 438 nm (Figure 4B) experiences no chromatographic interference, thus allowing for a highly selective determination of biotin (Figure 4 and Table I).

System III. This postcolumn reaction detection system for the HPLC determination of biotin is based on the use of a homogeneous non-competitive (compared to systems I and II) fluorophore-linked binding assay (Przyjazny, A.; Bachas, L. G. unpublished data). In particular, the system takes advantage of the enhancement of the fluorescence intensity of fluorescein-labeled avidin upon binding to biotin (29) or its derivatives (Barbarakis, M. S.; Bachas, L. G. unpublished data). In the instrumental setup described in Figure 1, the effluent stream from the HPLC column is mixed with a reagent stream containing avidin, labeled with fluorescein isothiocyanate, and the analyte concentration is determined by monitoring the analyte-induced increase in fluorescence at 520 nm (Figure 5). The method was validated by determining the biotin content in real samples (a liquid vitamin preparation and a

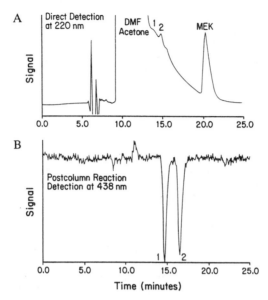

Figure 3. Chromatograms of 20 µL of a mixture containing 2.0 x 10^{-5} M each of (1) biotin and (2) biocytin and 0.07 M each of the interferents. (A) Direct UV detection at 220 nm, and (B) postcolumn reaction with fluorimetric detection at 438 nm. (Reproduced with permission from ref. 27. Copyright 1991 Elsevier.)

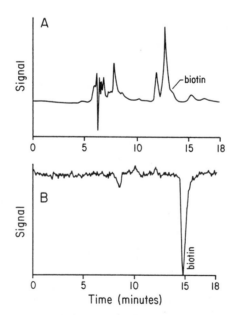

Figure 4. Chromatograms of 20 µL of a horse-feed supplement extract (A) with direct UV detection at 220 nm, and (B) with a postcolumn reaction detection scheme based on the competition of biotin and 2,6-ANS for the binding sites of avidin. (Reproduced with permission from ref. 27. Copyright 1991 Elsevier.)

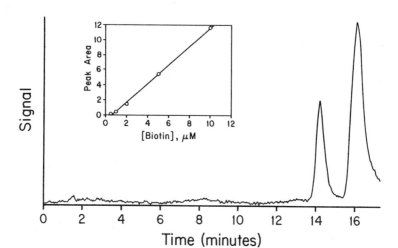

Figure 5. Chromatograms of 20 μL of a mixture containing 2.0 x 10⁻⁶ M each of biotin (t_R = 14.6 min) and biocytin (t_R = 16.6 min). The inset shows the calibration curve for biotin.

horse-feed supplement). For both samples, the postcolumn reaction system was much more selective than the direct UV detection at 220 nm. In addition, the amounts of biotin found in the real samples were within 2-3% of the amounts claimed by the manufacturers, which attests to the high accuracy of the system. The detection limits were 0.66 and 1.0 ng for biotin and biocytin, respectively, which are the lowest values reported in the literature for the HPLC determination of these analytes.

Merits of Systems I - III. Table II summarizes the detection limits and the cost per analysis for the three systems described above. A significant improvement in the detection limits achieved for biotin is shown, compared to the detection limit of 6 ng of biotin obtained using direct UV detection. In addition, it is noteworthy the fact that the system that achieves the lowest detection limits for biotin, exhibits the lowest cost per analysis as well (system III). This is a result of using a more sensitive binding assay, which allows for a lower reagent consumption in system III compared to systems I and II.

Table II. Merits of the Systems I - III

System	Detection limits for biotin (ng) [a]
Direct detection at 220 nm	6
I	6
II	2.4
III	0.66

[a] The given detection limits refer to a sample volume of 20 μL.

Future Prospects

The availability of several binding assay techniques for a variety of analytes (3), along with the fact that both antibodies and enzymes can function in the presence of small amounts of the typical organic modifiers used in liquid chromatography, enable us to conclude that the application of other binding assays in the development of on-line postcolumn reaction detection schemes for liquid chromatography would be quite feasible. For instance, the postcolumn reaction system III mentioned above can be extended to include other analytes as long as there exist fluorophore-labeled biological binders (e.g., antibodies, binding proteins, receptors, etc.) that undergo spectral changes as a function of the concentration of the analytes. Additional systems that are potentially useful are homogeneous fluorophore- or enzyme-linked competitive binding assays, where the binding of the ligand-specific binder to the fluorophore- or enzyme-labeled ligand alters the measured signal. Research along these lines is currently underway in our laboratory.

Acknowledgments. The research from the authors' laboratory was supported by a grant from the National Institutes of Health (Grant No. GM40510).

Literature Cited

1. *Affinity Techniques;* Jakoby, W. B.; Wilchek, M., Eds.; Methods in Enzymology; Academic Press: New York, 1974; Vol. 34.

2. Porath, J. *J. Chromatogr.* **1981**, 218, 241-259.
3. Gosling, J. P. *Clin. Chem.* **1990**, 36, 1408-1427.
4. Brown, P. R. *Anal. Chem.* **1990**, 62, 995A-1008A.
5. Erni, F. *J. Chromatogr.* **1990**, 507, 141-149.
6. Newton, P. *LC·GC* **1990**, 8, 706-714.
7. Rossi, T. M. *J. Pharm. Biomed. Anal.* **1990**, 8, 469-476.
8. Säwe, J. *Adv. Pain Res. Ther.* **1986**, 8, 45-55.
9. O'Connor, E. F.; Cheng, S. W. T.; North, W. G. *J. Chromatogr.* **1989**, 491, 240-247.
10. Santos, S. R. C. J.; Kirch, W.; Ohnhaus, E. E. *J. Chromatogr.* **1987**, 419, 155-164.
11. Naline, E.; Flouvat, B.; Pays, M. *J. Chromatogr.* **1987**, 419, 177-189.
12. Young, T. L.; Habraken, Y.; Ludlum, D. B.; Santella, R. M. *Carcinogenesis* **1990**, 11, 1685-1689.
13. Stone, J. A.; Soldin, S. J. *Clin. Chem.* **1988**, 34, 2547-2551.
14. Twitchett, P. J.; Fletcher, S. M.; Sullivan, A. T.; Moffat, A. C. *J. Chromatogr.* **1978**, 150, 73-84.
15. Williams, P. L.; Moffat, A. C. *J. Chromatogr.* **1978**, 155, 273-283.
16. Moffat, A. C. *Acta Pharm. Suec.* **1978**, 15, 482.
17. O'Keeffe, M.; Hopkins, J. P. *J. Chromatogr.* **1989**, 489, 199-204.
18. Rapp, M.; Meyer, H. H. D. *J. Chromatogr.* **1989**, 489, 181-189.
19. Diamandis, E. P.; Christopoulos, T. K. *BioTechniques* **1991**, 10, 646-648.
20. Gelpi, E.; Ramis, I.; Hotter, G.; Bioque, G.; Bulbena, O.; Rosello, J. *J. Chromatogr.* **1989**, 492, 223-250.
21. De Alwis, W. U.; Wilson, G. S. *Anal. Chem.* **1985**, 57, 2754-2756.
22. De Alwis, W. U.; Wilson, G. S. *Anal. Chem.* **1987**, 59, 2786-2789.
23. Heineman, W. R.; Halsall, H. B.; Wehmeyer, K. R.; Doyle, M. J.; Wright, D. S. *Methods Biochem. Anal.* **1987**, 32, 345-393.
24. Przyjazny, A.; Kjellström, T. L.; Bachas, L. G. *Anal. Chem.* **1990**, 62, 2536-2540.
25. Green, N. M. *Biochem. J.* **1965**, 94, 23c-24c.
26. Green, N. M. *Adv. Protein Chem.* **1975**, 29, 85-133.
27. Przyjazny, A.; Bachas, L. G. *Anal. Chim. Acta* **1991**, 246, 103-112.
28. Mock, D. M.; Langford, G.; Dubois, D.; Criscimagna, N.; Horowitz, P. *Anal. Biochem.* **1985**, 151, 178-181.
29. Al-Hakiem, M. H. H.; Landon, J.; Smith, D. S.; Nargessi, R. D. *Anal. Biochem.* **1981**, 116, 264-267.

RECEIVED February 24, 1992

Chapter 12

Determination of a New Oral Cephalosporin in Biological Fluids

Liquid Chromatographic, Ultraviolet, and Mass Spectrometric Methods

Hwai-Tzong Pan[1], Pramila Kumari, and Chin-Chung Lin

Department of Drug Metabolism and Pharmacokinetics, Schering–Plough Research, Bloomfield, NJ 07003

A LC/UV method has been previously developed to determine ceftibuten (SCH 39720), a new oral cephalosporin, in human plasma. The method, used routinely for bioavailability/ pharmacokinetic studies, requires minimal sample preparation and a small sample volume (100 µl). It can reliably determine concentrations of ceftibuten as low as 1 µg/ml. However, the method was found not to be suitable for the determination of ceftibuten concentrations in tracheal and bronchial secretion or middle ear fluid. These fluids contain matrix interferences and their small sample volume usually does not exceed 30 µl. We therefore developed a LC/LC/UV method which successfully resolves ceftibuten from matrix interference and requires a minimal sample volume of only 20 µl. The assay has very good sensitivity of 0.05 µg/ml. In addition, a LC/LC/MS method will be discussed. This method has been developed to determine the drug in sputum samples, where the other methods proved unsuccessful.

Ceftibuten; (+)-(6R,7R)-7-[(Z)-2-(2-Amino-4-thiazolyl)-4-carboxycrotonamido]-8-oxo-5-thia-1-azabicyclo[4.2.0]-oct-2-ene-2-carboxylic acid (SCH 39720), Figure 1, is a new oral cephalosporin antibiotic which was discovered by Shionogi Research Laboratories, Ltd, Tokyo, Japan. It is highly active against a variety of Gram-negative bacteria; E. coli., Klebsiella, Proteus, and H. influenzae and moderately active against the Gram-positive bacteria; Enterobacter, Citrobacter, Serratia, and S. pyogenes, respectively.

A conventional high performance liquid chromatography

[1]Current address: Department of Pharmacology, National Yang-Ming Medical College, Shih-Pai, Taipei 11221, Taiwan

0097–6156/92/0512–0144$06.00/0
© 1992 American Chemical Society

(HPLC) method with ultraviolet detection (UV) has been developed by Lim et al., [3], for determining concentrations of Ceftibuten in plasma in normal volunteers by direct sample injection. Unfortunately, this method did not provide enough selectivity to assay tracheal secretions, middle-ear fluids, and sputum samples from patients. Although samples pretreated by liquid-liquid or solid-phase extration can yield cleaner sample matrices, they are labor-intensive, time comsuming, and provide less recovery. The small volume of middle ear fluid samples from pediatric volunteers are extremely difficult to handle. Column-switching HPLC techniques have gained increasing attention because it not only can offer more resolution but can also circumvent the labor intensive extraction procedure, the potential for sample decomposition at room temperature and cross contamination over time [1, 6-9].

Column-switching HPLC analysis can also accomodate large sample injection volumes and provide trace enrichment of the analyte resulting in narrow peaks, thereby providing better sensitivity. A column-switching HPLC assay combined with a heart-cutting technique using either ultraviolet detection (LC/LC/UV) or thermospray mass spectrometry (LC/LC/TSP-MS) was subsequently developed. The LC/LC/UV assay reported here is for middle ear fluid samples analysis [4]. The LC/LC/TSP-MS was selected for quantitative analysis of ceftibuten in sputum samples [2, 5, 10]. The overall assay characteristics of this procedure including the TSP-mass spectrometric parameters, data acquisition, method validation such as: linearity, sensitivity, specificity, accuracy, and precision are discussed.

Materials and Methods

Reagents. HPLC grade acetonitrile was obtained from Fisher Scientific (Pittsburgh, PA). Deionized water was collected from a Milli-Q water purification system (Millipore Corp., Milford, MA). Ammonium Acetate of ACS grade was from Sigma Chemical Company (St. Louis, MO). The ceftibuten capsules were supplied by Schering-Plough Corporation (Kenilworth, NJ). Polyethylene glycol, average M.W. 300 (PEG 300), was from Aldrich Chemical Company (Milwaukee, WI).

Apparatus. The column-switching chromatographic system consisted of two solvent delivery pumps. Pump A (Waters Model 590) was used for the delivery of "weak" mobile phase for analyte clean up and preconcentration and pump B (Waters 600-MS Multisolvent Delivery System) was used for the delivery of the "strong" mobile phase. Two standard HPLC six-port valves were used (Rheodyne Model 7000). One was for direct sample injection and the second was used to switch the analyte containing fraction eluting from the first-preparative column into the second-analytical column for quantitation. The UV detector was a Waters Lambda-Max

(Model 481) operated at 263 nm and the chromatograms were
recorded on a Waters SE-120 strip chart recorder.   The
instrument used for TSP mass spectrometry was a model 201
LC/MS system (Vestec, Houston, Texas) controlled by a
Teknivent vector one data system (Teknivent, St. Louis,
Missouri). LC/LC/TSP-MS was performed in the positive ion-
selected ion monitoring mode with the filment off [5]. The
thermospray optimized conditions were as follows: T1, 170
^{0}C; T2, 225 ^{0}C (probe tip temperature); T3, 312 ^{0}C (block
temperature); T4, 270 ^{0}C (vapor temperature). The schematic
diagram of the equipment setup is shown in Figure 2. There
is a two-way valve after column B in order to choose either
UV or MS detection.

Chromatography. The first preparative column used on-line
for   sample   extraction   and   clean-up   was   a   Waters
microbondapak phenyl packing in a 3.9 mm i.d. X 150 mm
stainless steel column (particle size 10 μm). The second
analytical column was a microbondapak phenyl packing in a
3.9 mm i.d. X 300 mm stainless steel column (partical size
10 μm).   The weak ionic mobile phase used in the first
column was 0.1 M ammonium acetate (pH=6.5) and the strong
ionic mobile phase used in the second column was 2%
acetonitrile in 0.1 M of ammonium acetate (pH=6.5).   The
use of ammonium acetate buffer is needed for the
thermospray ionization process.   The flow rate in both
columns was 1 ml/min and the time window to elute the
analyte, ceftibuten, from the first preparative column into
the second analytical column was 4 to 6 minutes. Duplicate
injections of a 1 μg/ml ceftibuten standard into the
LC/LC/UV system showed that the overall retention time
through the two column system was about 13 minutes, Figure
3.   A guard column was inserted before the first column in
order to prevent lipoproteins from depositing on the head
of the column and was changed every day. The first column
was flushed with 2% acetonitrile in 0.1 M ammonium acetate
overnight in order to elute any matrix residues adsorbed on
the column.

Sample Processing. All samples were stored in a freezer at
-70°C. On the day of analysis, the samples were thawed at
room temperature, the weight of the sample tube and sample
was recorded (W1).   A 200 μl aliquot of 0.1 M ammonium
acetate solution was added into the sample and mixed well
by vortex action mixing.   The diluted samples were then
centrifuged for 20 minutes at 3000 rpm.   The supernatant
was transferred into clean vials using a micro pipette. A
one hundred microliter (100 μl) aliquot of the supernatant
from a middle ear fluid sample was injected directly into
the LC/LC/UV system, while a similar one hundred microliter
aliquot of the supernatant from a sputum sample was
injected for LC/LC/TSP-MS analysis.   The sample tube was
thoroughly washed and air dried after sample assay.   The
weight of empty sample tube after drying was recorded (W2).
The net weight (mg) of the sample was taken as (W1 - W2).

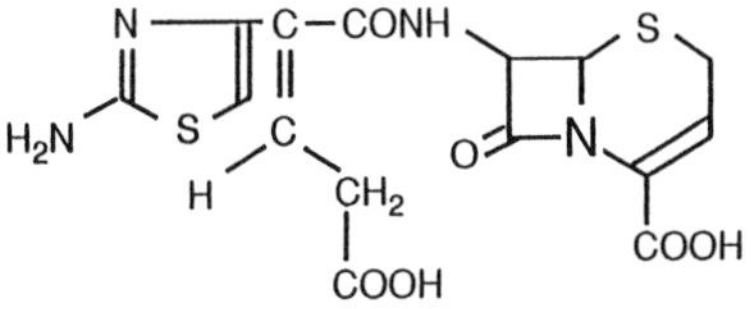

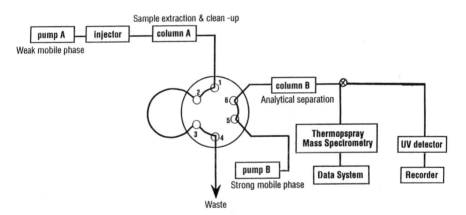

Figure 1. Chemical structure of ceftibuten (SCH 39720).

Sample extraction & clean -up

Figure 2. Schematic diagram of equipment for the LC/LC/UV and LC/LC/TSP-MS system.

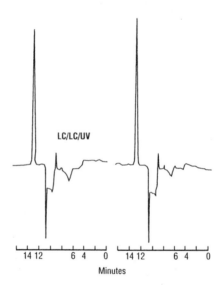

Figure 3. Typical chromatograms of ceftibuten standards (1 μg/ml) obtained using the LC/LC/UV system. The overall retention time of ceftibuten was 13 min. (Reproduced with permission from reference 4, 1992 American Pharmaceutical Association)

Data Acquisition. All the calibration standards were prepared in 0.1 M ammonium acetate buffer because of insufficient supply of the respective control human biological fluids. After analysis, the concentration (C) in the diluted sample was calculated using the calibration curve. The calibration curve was determined by the regression line defining the linear relationship between peak area response of each ceftibuten standard vs theoretical concentration: y = mx + b, where x = peak area of ceftibuten, y = concentration of ceftibuten (µg/ml), m = slope of the fitted line, b = the y intercept of the regression line. No weighting factors were used to calculate the regression line. The real concentration of ceftibuten (C_{real}) for the samples were expressed by C * [200+(W1-W2)]/(W1-W2)] by assuming the density of fluids as 1 g/ml.

Results and Discussion

Sensitivity and Linearity. For LC/LC/UV, the limit of quantitation (LOQ) was established at 0.10 µg/ml. Duplicate chromatograms at the limit of detection of 0.05 µg/ml are shown in Figure 4. Injection of ceftibuten standards ranging from 0.10 µg/ml to 5 µg/ml (n=15), established that the correlation coefficient of the calibration curve was 0.996.

For LC/LC/TSP-MS, the LOQ was validated at 0.50 µg/ml and a typical thermospray ion chromatogram at the LOQ is illustrated in Figure 5. The LC/LC/TSP-MS calibration curves were linear in the concentration range of 0.50 to 10.00 µg/ml (n=35) with a correlation coefficient of 0.986. The LC/LC/TSP-MS assay offers only better selectivity than the LC/LC/UV method with no advantage in sensitivty.

Assay Specificity. For LC/LC/UV method, no interfering peaks were found in the clinical pre-dose (0 hr) middle ear fluid sample at the overall retention time of ceftibuten (13 minutes), Figure 6A. A typical chromatogram from a four-hour post dose middle ear fluid sample is shown in Figure 6B.

The LC/LC/UV method was not sufficiently selective for sputum samples due to an interfering peak in the retention area of ceftibuten (R_t=13 minutes) in two pre-dose sputum samples from different patients (Figure 7). However, the thermospray mass chromatogram from a pre-dose control sputum sample analyzed by LC/LC/TSP-MS, Figure 8, shows that no interfering peaks were found in the retention area of ceftibuten.

The only potential interference of ceftibuten for LC/LC/TSP-MS methods would be the trans-isomer (metabolite), which however has a longer retention time (6 to 8 min) in the first preparative column and would not be eluted in the 4 to 6 minute segment containing ceftibuten switched into the second analytical column (Figure 9).

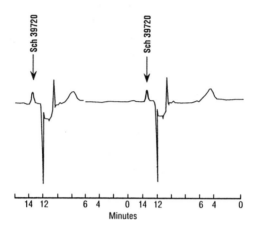

Figure 4.   Duplicate chromatograms at the detection limit (0.05 µg/ml) for the LC/LC/UV system. (Reproduced with permission from reference 4, 1992 American Pharmaceutical Association)

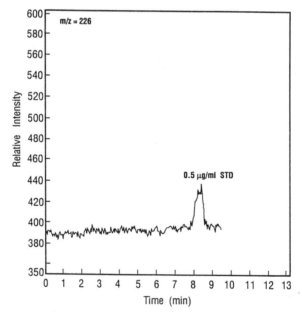

Figure 5.   LC/LC/TSP-MS analysis at the limit of quantitation (0.50 µg/ml). (Reproduced with permission from reference 5, 1992 American Pharmaceutical Association)

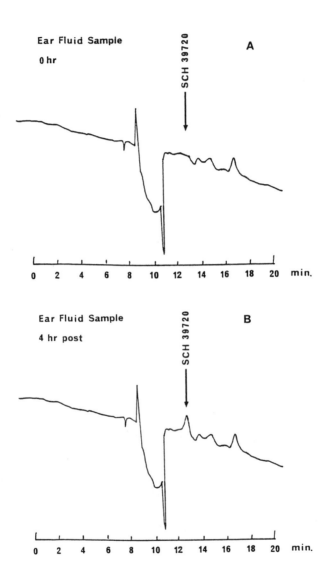

Figure 6. (A) Chromatogram from a zero-hour post dose middle ear fluid sample and (B) from a four-hour post dose middle-ear fluid sample using LC/LC/UV system.

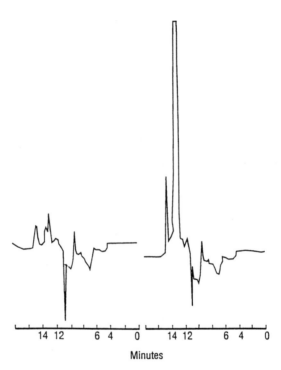

Figure 7. The chromatograms of two 0 hr sputum samples detected by LC/LC/UV method. (Reproduced with permission from reference 5, 1992 American Pharmaceutical Association)

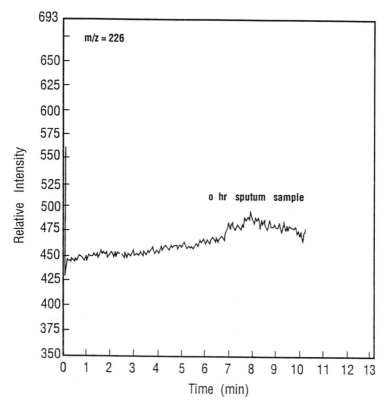

Figure 8.   LC/LC/TSP-MS analysis of 0 hr drug free sputum sample, showing no interfering peaks in the retention area of ceftibuten. (Reproduced with permission from reference 5, 1992 American Pharmaceutical Association)

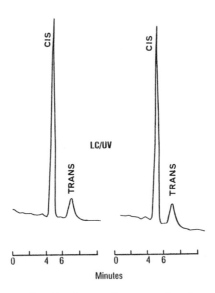

Figure 9. A typical chromatogram is demonstrated for the separation of a mixture of ceftibuten and its major metabolite (trans) by the first sample preparative column. (Reproduced with permission from reference 4, 1992 American Pharmaceutical Association)

Accuracy and Precision. For the LC/LC/UV assay, the precision and accuracy were demonstrated by the injection of 15 ceftibuten standards. The coefficient of variation ranged from 0 to 13 and the %bias ranged from 0 to 9 (Table I).

For the LC/LC/TSP-MS method, the assay was validated by injection of 35 ceftibuten standards, and shown to be precise with a variability (%CV) of 20% or less, and the overall accuracy was 96.4 ± 4.4 %, mean value of (100 - Bias)%, Table II.

Clinical Applications. The clinical utility of these methods were demonstrated by (1) determining the middle ear fluid concentrations of ceftibuten following a 10 day course of ceftibuten suspension, 9 mg/kg/day, in a single daily dose (maximun dose of 400 mg a day) in thirteen patients with otitis media using the LC/LC/UV method (Figure 10), and (2) determining the sputum concentrations of ceftibuten after a two x 200 mg dose of ceftibuten capsules (400 mg) in twelve patients with tracheostomy or laryngectomy using the LC/LC/TSP-MS method (Figure 11), respectively. Both the LC/LC/UV and LC/LC/TSP-MS system were shown to be sensitive, reliable, and selective for the quantitation of ceftibuten in different biological fluids.

Table I Precision and accuracy of the LC/LC/UV assay for ceftibuten

Theoretical concentration (μg/ml)	Observed concentration (μg/ml)	%CV	%Bias
0.10	0.10 (n=3)	12	0
0.25	0.25 (n=2)	6	0
0.50	0.51 (n=5)	11	2
1.00	0.91 (n=4)	13	9
5.00	4.97 (n=1)	0	1

Table II Precision and accuracy of the LC/LC/TSP-MS assay for ceftibuten

Theoretical concentration (μg/ml)	Observed concentration (μg/ml)	%CV	%Bias
0.50	0.46 (n=9)	17.4	8.0
1.00	0.92 (n=9)	19.9	8.0
2.50	2.51 (n=7)	12.1	0.4
5.00	5.09 (n=9)	4.2	1.8
10.00	9.60 (n=1)	0.1	4.0

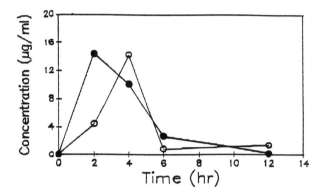

Figure 10. Mean concentration-time curves for ceftibuten in middle ear fluid and plasma in pediatric patients following multiple oral administration of ceftibuten (9 mg/kg/day) once daily for 3 days. (Filled circle: plasma; Open circle: middle ear fluid)

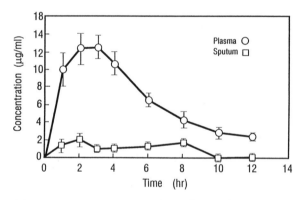

Figure 11. Mean concentration-time curves for ceftibuten in plasma and sputum samples after a single 400 mg oral dose of ceftibuten in patients undergoing tracheostomy or laryngectomy. (Reproduced with permission from reference 5, 1992 American Pharmaceutical Association)

References

1.   S.A. Berkowitz, **Anal. Biochem.** 1987, 164, 254–260.
2.   W.J. Blanchflower and D.G. Kennedy, **Biomed. Environ. Mass Spectrom.**, 18 (1989) 935–936.
3.   J.M. Lim, Hong Kim, and C.C. Lin, **Antimicrobial Agents and Chemotherapy** 1991 (submitted).
4.   H.T. Pan, P. Kumari, J. Lim, and C.C. Lin, J. **Pharm. Sci.**, 81 (1992) 1–4.
5.   H.T. Pan, P. Kumari, J.A.F. de Silva, and C.C. Lin, J. **Pharm. Sci.**, 1992 (in press).
6.   J.V. Posluszny and R. Weinberger, **Anal. Chem.** 1988, 60, 1953–1958.
7.   J.V. Posluszny and R. Weinberger, J. **Chromatogr.** 1990, 507, 267–276.
8.   D.A. Roston and R. Wijayaratne, **Anal. Chem.** 1988, 60, 950–958.
9.   W.J. Sonnefeld, W.H. Zoller, W.E. May, and S.A. Wise, **Anal. Chem.** 1982, 54, 723–727.
10.   R.D. Voyksner, J.T. Bursey, and E.D. Pellizzari, **Anal. Chem.**, 56 (1984) 1507–1514.

RECEIVED June 3, 1992

Chapter 13

Drugs and Metabolites in Biological Fluids
Automated Analysis by Laboratory Robotics

Linda A. Brunner

Drug Development Department, Pharmaceuticals Division, Ciba–Geigy
Corporation, Ardsley, NY 10502

A laboratory robotic system designed specifically for automating the
sample preparation procedure and chromatographic analysis of drugs
and their metabolites in biological fluids has been installed and
validated.   The system, directly interfaced to a high per-formance
liquid chromatograph and data station, has been designed for routine
use in pharmacokinetic and disposition studies involving both
experimental and marketed non-steroidal anti-inflammatory drugs, as
well as several anti-epileptic compounds. PyTechnology architecture,
combining new advances in both software and hardware, is utilized to
perform all the necessary laboratory operations for the separation of
the compounds of interest from the biological matrix and their
subsequent chromatographic analysis. Uniform sample treatment and
rapid data collection, with minimal human intervention, are achieved
through serial sample preparation. The automated procedure is carried
out reliably and reproducibly during continuous unattended operation
of the system. Validation of the entire automated sample preparation
procedure as well as the individual system components was completed
over the course of several weeks. The validation data, in addition to
the application of the procedure to the analysis of the different
compounds in clinical samples, will be demonstrated. Comparisons
to the initial manual methods as well as other, older custom robotic
systems already in the laboratory will be presented.

A PyTechnology laboratory robotic system designed specifically for automating the
extraction and analysis of drugs and their metabolites from biological fluids has
been installed in our laboratories. System validation was initially performed using
prinomide and its para-hydroxy metabolite as an example of an anti-inflammatory
drug candidate (1). A earlier automated procedure for these compounds (2) was
modified in order to accomodate the new hardware and software used by this new

0097–6156/92/0512–0158$06.00/0

applications system. Subsequently, methodology for a marketed anti-epileptic drug, Tegretol (carbamazepine, CBZ), and its 10,11-epoxide metabolite (CBZE) has also been developed and validated on this new robotic system (Brunner, L. A. etal. *Lab. Auto. Rob.*, in press). Previously installed custom robotic systems designed for trace level determinations of drugs and metabolites in biological fluids were used to develop and validate methodology for a marketed anti-inflammatory drug, Voltaren (diclofenac sodium), (*3*) as well as an anti-epileptic drug candidate presently under clinical development (Brunner, L. A. etal. *Biomed. Chrom.*, in press). The validation data, in addition to applications for the analysis of each of the compounds in clinical samples, will be demonstrated. Comparisons to inital manual methods will also be presented. The robotic systems provide unattended sample preparation and analysis, as well as automated retrieval and computation of the chromatographic results.

Experimental

Chemicals and Reagents. All of the compounds (active ingredients, drug candidates) and their respective metabolites as well as internal standards were supplied by Ciba-Geigy Corporation (Summit, NJ, USA), except for the internal standard for carbamazepine, cyheptamide (CHA, 10-ml vial at 1 mg/ml in methanol), which was purchased from Supelco, Inc., Bellefonte, PA, USA. The HPLC-grade solvents (methanol, water, acetonitrile, methyl t-butyl ether, isopropyl alcohol, hexane, ethyl acetate and dichloromethane) used for the analyses were obtained from Burdick & Jackson, Muskegon, MI, USA. Sodium hydroxide pellets, potassium dihydrogen phosphate, potassium phosphate (monobasic), sodium citrate, citric acid, 25% tetrabutyl-ammonium hydroxide in methanol (TBAH), monobasic and dibasic sodium phosphate and phosphoric acid (85%), all reagent grade, were purchased from J.T. Baker Chemical Company, Phillipsburg, NJ, USA. Control (drug free) human plasma from heparinized blood, used for the standard curves and quality control samples, was obtained from Biological Specialty Corporation, Landsdale, PA, USA.

Materials/Labware. All disposable culture tubes, pipet tips and screw-cap tubes utilized on the robotic system for the sample preparation, were purchased from Baxter Healthcare Corporation, McGaw Park, IL, USA. The teflon-lined caps used on the systems were purchased from either Sunbrokers, Inc., Wilmington, NC, USA or Wheaton, Milville, NJ, USA.

Preparation of Reagents and Standard Solutions. All mobile phase solutions were prepared fresh daily, filtered through a 0.45-μm Millipore GVWP mem-brane filter (Millipore Corp., Bedford, MA, USA) and degassed under vacuum prior to use.

Stock solutions (1.0 mg/ml) of each of the compounds of interest (drugs and metabolites) were prepared in methanol, and the necessary dilutions were made with water to give the appropriate spiking solutions.

Stock solutions of the respective internal standards were also prepared in methanol, and diluted with water to give the internal standard spiking solutions.

All of the solutions were stored at about 4°C and brought to room temperature prior to use.

Calibration standards (standard curve samples) were prepared in duplicate, on a daily basis, by spiking 3.0 ml of human control plasma with the appropriate solutions of the respective compounds of interest and any metabolites. Spiked plasma pools were prepared for the quality control samples by diluting the appropriate volumes of the stock or spiking solutions with control human plasma to result in concentrations at the lower and upper ends of, as well as in between, the ranges studied for each of the drugs and their respective metabolites. These quality control samples were prepared in advance and were stored as 3.0-ml aliquots at about -20°C until analyzed.

Instrumentation. A Zymate II and a PyTechnology laboratory automation system (Zymark Corporation, Hopkinton, MA, USA) were used for sample preparation. The robotic systems were programmed through the use of Easylab software (Zymark Corp.). The pneumatic actuation of several of the robotic modules was controlled by a Jun-Air compressor (Jun-Air, Inc., Racine, WI, USA). The entire system was powered through an uninter-ruptable power supply (UPS) and voltage regulator (SOLA Electric, Elk Grove Vlg, IL, USA). The Zymate II system (Figure 1) is set-up on a 4 x 6 foot movable cart that can be strategically situated in the laboratory to best accomodate needed periphery, such as chromatographic equipment and ventilation systems. All locations on the table, as well as the final serialized sample preparation procedure/method, are programmed into the system controller by the end-user. The PyTechnology robotic system (Figure 2) is located on a 5 x 7 foot movable cart. Each module (Pysection) comes with pre-programmed software, leaving only the location at which the section was installed on the table for the end-user to program into the controller. The top-level method program also has to be built and programmed by the end-user, incorporating variables (ie. volumes, times, etc.) specific to the individual procedures. The laboratory robotic systems cost approximately $75,000 to $125,000, depending on the complexity of the methods to be automated. Routine monthly maintenance ensures system reliability and minimum down-time. When calculating the systems' speed of operation, one must compare a technician's 8-hr day to a 24-hr period for the robot. For our applications, a technician can prepare samples faster than an automated system on a one-to-one basis; however, when the automated system is left running unattended, in the serial mode, over a 24-hr period, the robot can perform 1.5 times the amount of work than a technician in an 8-hr workday.

Analyses were performed using an HPLC system consisting of: two Model 590 programmable solvent delivery modules and a Model 680 gradient controller (Waters Associates, Milford, MA, USA), or a Model 8800 ternary gradient pump (Spectra-Physics, San Jose, CA, USA); a C_8 or C_{18} analytical column (dependent on the methodology developed for each specific compound of interest) in series with a 0.5μm in-line filter (Upchurch Scientific, Inc., Oak Harbor, WA, USA) and a 3-10μm guard column cartridge and holder (Brownlee Labs, Applied Biosystems, Santa Clara, CA, USA or Waters Assoc.); and, a Spectraflow/Kratos Model 783A UV detector (Applied Biosystems, Inc.) set to the specific wavelength for each compound analyzed. Chromatographic separations were performed in the reversed-

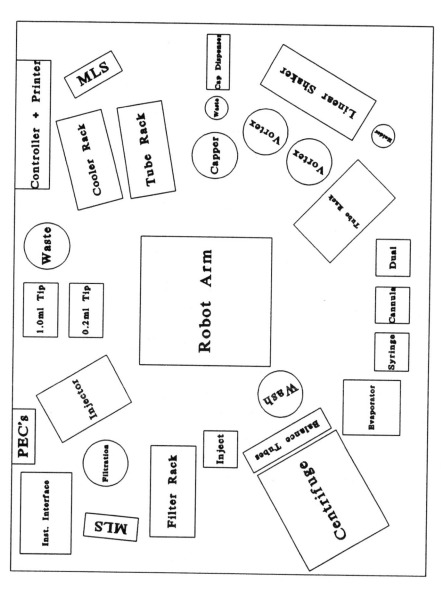

Figure 1. Zymate II laboratory automation (robotic) system layout.

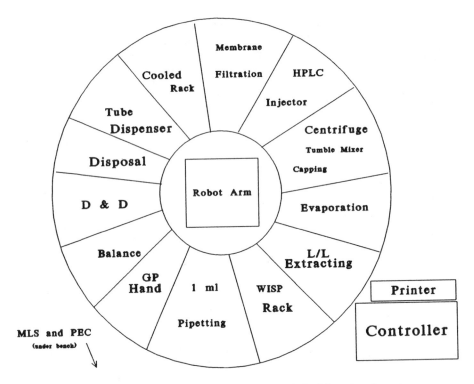

Figure 2. PyTechnology laboratory automation (robotic) system layout.

phase mode using individual (specific) mobile phase solutions (prepared v/v) at temperatures of 25-40°C, under pressures of 110-200 bar, with the flow rates of the mobile phase set at 1.0-1.5 ml/min (all parameters dependent on the individual methods developed for each of the compounds).

Peak areas for the drugs, metabolites and internal standards were measured using a Model 970 (256K memory) dual channel interface and a TurboChrom chromatography workstation (PE Nelson Systems, Inc., Cupertino, CA, USA). The chromatographic data were processed for peak area ratios of the drug(s) and/or metabolite(s) to internal standard, followed by weighted (1/peak area ratio) linear least squares regression analysis using TurboChrom 2700 software (PE Nelson Systems, Inc.).

Sample Preparation and Analysis. The robotic systems prepare the samples similar to the manual sample treatment procedure developed for each compound and its respective metabolite, with minor modifications being implemented that are necessary to accomodate robotic modules and peripherals. Figure 3 shows the general flow chart of the sample preparation procedures used for the analysis of all of the compounds reported. However, for the robotic method, the sample preparation is carried out serially, rather than in batch mode, as is done manually. Therefore, when fully operational, the robot works on four or five samples simultaneously, all of which are at various stages of completion. For the chromatographic analysis, a 10 to 50-μl aliquot was injected onto the respective chromatographic system, depending on the compound being analyzed.

Data acquisition was initiated automatically from the robotic system through the instrument interface module. Calibration curves were generated by the TurboChrom workstation (PE Nelson Systems, Inc.) and were represented by a plot of peak area ratios of the drugs and metabolites to internal standard versus the concentrations of the respective calibration standards. Quantification of quality control samples and unknown (clinical) samples was obtained by interpolation from the equations of the regression lines of the respective calibration curve.

Results

System Validation. The individual components of the sample preparation procedures such as pipetting, liquid dispensing, etc., as well as each entire method was tested and validated so as to establish the systems' integrity, reliability and reproducibility (Table I).

Specificity. The specificity of each method was demonstrated by the lack of interferences observed at the retention times of all the compounds of interest (drugs, metabolites and internal standards) in control human plasma from heparinized blood of volunteers ($N \geq 6$) receiving no medication.

Limit of Quantification/Detection. The limit of quantification (LOQ), defined as the lowest concentration where acceptable accuracy (mean relative recovery of $100 \pm 15\%$) and precision (CV $\leq 15\%$) are obtained, was established for each compound under its specific method conditions. Using the individual sample preparation procedures and a signal to noise ratio of 4:1, the minimum detectable amount of each compound of interest was also determined.

Table I. Validation of the PyTechnology Robotic System (N=50 to 400)

Laboratory Operation	CV (%)
Sample Volume	0.22 - 0.56
Int Std Volume	1.1 - 1.8
Injection Volume	3.2 - 7.6
Method/System	5.0 - 10.8
Quality Control	7.0 - 15.0

Recovery from Plasma. Extraction efficiencies (absolute recoveries) for each drug, metabolite and internal standard were determined by comparing peak areas of the analytes from extracted plasma standards to those from chromatographic standard solutions prepared in mobile phase at the equivalent concentrations and chromatographed directly.

Linearity. Linearity data were obtained following the analysis of spiked plasma with all the described compounds of interest. Each method was validated over a specific dynamic concentration range for each compound (drug/metabolite). Typical response vs. concentration plots (Figures 4-7) of these data indicated good correlations to the linear regression models used for each of the individual methods.

Accuracy and Precision. Relative recovery data for the quality control samples were used to assess the overall accuracy of each of the methods. The reproducibility of the recovery data was used to determine the precision of the methods. The intra- (N=4-5) and inter-day (N=3-4) accuracy and precision data are expressed as mean percent found and coefficient of variation (CV), respectively, and are shown in Tables II-V. The inter-day values were calculated using all the determinations at the indicated concentrations.

The overall mean relative recovery and coefficients of variation were good for each of the compounds of interest, as indicated in the corresponding tables. Comparison of the data obtained robotically with those obtained for the respective manual methods showed similar results.

Discussion

This robotic technology was implemented to automate the preparation and analysis of human plasma samples for drug-level determinations. Utilization of a laboratory robotic system eliminates the tedium of the sample preparation

ALIQUOT PLASMA
↓
ADD INTERNAL STANDARD
↓
ADJUST pH RANGE
↓
VORTEX (MIX)
↓
ADD EXTRACTING SOLVENT
↓
CAP
↓
CONDITION (SHAKE/ROTATE)
↓
CENTRIFUGE
↓
TRANSFER ORGANIC LAYER
↓
EVAPORATE
↓
RECONSTITUTE
↓
INJECT ONTO HPLC SYSTEM
↓
COLLECT DATA

Figure 3. Flow chart of the sample preparation procedure.

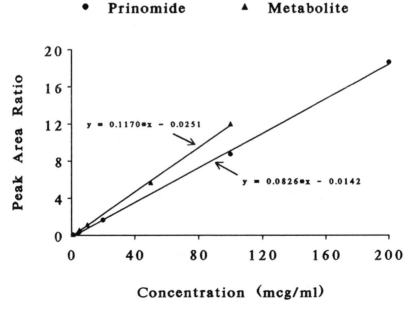

Figure 4. Typical calibration curve of the drug (prinomide) and its para-hydroxy metabolite in human plasma generated on the PyTechnology robot.

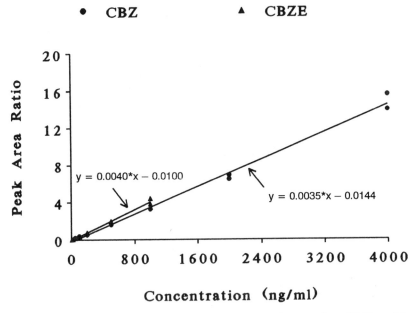

Figure 5. Typical calibration curve of the drug (carbamazepine, CBZ) and its 10,11-epoxide metabolite (CBZE) in human plasma generated on the PyTechnology robot.

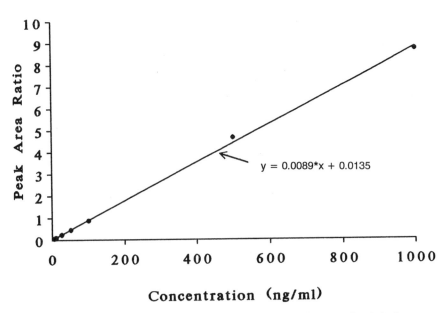

Figure 6. Typical calibration curve of the drug (diclofenac sodium) in human plasma generated on the Zymate II custom robot.

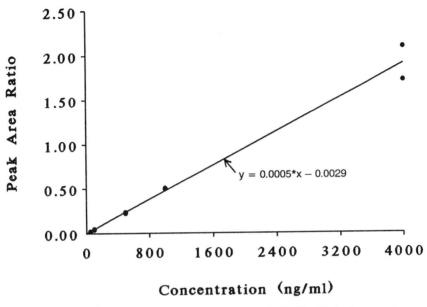

Figure 7. Typical calibration curve of the drug (CGP 33101) in human plasma generated on the Zymate II custom robot.

Table II. Intra- and Inter-Day Accuracy and Precision Data for Determination of Prinomide and its Para-hydroxy Metabolite Concentrations in Plasma on the PyTechnology Robot

Compound	Added Amount (μg/ml)	Day 1	Intra-Day Day 2	Day 3	Inter-Day
			Mean Percent Relative Recovery (% CV)		
Prinomide	1.0	104 (8.0)	116 (1.5)	97.3 (5.1)	104 (10.8)
	5.0	88.6 (6.9)	84.4 (7.8)	102 (3.1)	97.9 (5.0)
	20	95.6 (6.3)	92.4 (9.7)	110 (8.9)	100 (7.8)
	100	89.6 (2.0)	93.5 (9.0)	105 (6.4)	99.9 (9.4)
	200	102 (5.3)	97.9 (5.8)	96.3 (5.5)	100 (5.2)
Metabolite	1.0	112 (4.3)	87.0 (1.3)	92.7 (1.8)	100 (10.2)
	5.0	98.5 (3.9)	97.7 (5.0)	116 (7.0)	101 (7.2)
	10	96.8 (2.7)	102 (5.9)	102 (6.7)	99.4 (9.2)
	50	85.8 (9.4)	95.9 (7.1)	105 (6.3)	97.2 (10.5)
	100	106 (3.1)	92.0 (6.0)	89.3 (3.7)	101 (6.3)

Table III. Intra- and Inter-Day Accuracy and Precision Data for Determination of Carbamazepine and its 10,11-Epoxide Metabolite Concentrations in Plasma on the PyTechnology Robot

Compound	Added Amount (ng/ml)	Mean Percent Relative Recovery (% CV) Intra-Day			Inter-Day
		Day 1	Day 2	Day 3	
CBZ	50	113 (9.6)	116 (5.8)	85.8 (2.6)	96.2 (11.0)
	200	90.3 (9.0)	100 (9.1)	93.2 (3.8)	95.4 (7.8)
	1000	91.2 (6.3)	94.8 (1.6)	100 (1.4)	98.0 (4.8)
	4000	101 (2.2)	102 (13.0)	104 (1.1)	102 (6.0)
CBZE	10	105 (3.7)	103 (1.3)	83.3 (10.0)	96.2 (11.0)
	50	96.4 (5.4)	94.4 (3.8)	124 (3.6)	106 (12.0)
	500	93.0 (5.6)	95.9 (3.8)	100 (2.0)	97.7 (5.1)
	1000	98.5 (4.7)	101 (13.0)	104 (1.7)	101 (6.7)

Overall Mean Relative Recovery = 98.5% (CBZ) and 100% (CBZE)

Table IV. Intra- and Inter-Day Accuracy and Precision Data for Determination of Diclofenac Sodium Concentrations in Plasma on the Zymate II Custom Robot

Added Concentration (ng/ml)	Mean Percent Relative Recovery (% CV) Intra-Day			Inter-Day
	Day 1	Day 2	Day 3	
5.0	118 (5.8)	110 (10.3)	99.1 (7.6)	109 (10.3)
15	101 (3.4)	103 (2.6)	95.6 (1.9)	99.5 (3.9)
800	99.6 (4.9)	95.9 (0.7)	92.7 (5.4)	94.8 (3.9)
1000	95.5 (0.7)	102 (0.5)	108 (1.5)	100 (5.4)

Overall Mean Relative Recovery = 99.8%

Table V. Intra- and Inter-Day Accuracy and Precision Data for Determination of CGP 33101 Concentrations in Quality Control Samples on the Zymate II Custom Robot

Added Concentration (ng/ml)	Mean Percent Relative Recovery (% CV) Intra-Day			Inter-Day
	Day 1	Day 2	Day 3	
50	110.1 (11.3)	108.8 (3.2)	114.1 (7.1)	110.8 (7.7)
100	106.4 (3.7)	99.7 (3.2)	104.0 (6.0)	103.6 (5.1)
1000	104.5 (5.2)	94.6 (4.3)	100.6 (3.7)	99.9 (6.0)
4000	96.7 (1.4)	90.4 (1.4)	98.0 (3.7)	95.4 (4.4)

Overall Mean Relative Recovery = 102.7%

procedure, improves safety by ensuring minimal exposure to biological samples, and frees the analyst to perform other duties, by leaving only the replenishment of disposables (tubes, caps and pipet tips) and filling of reagent reservoirs. During continuous unattended operation, the robot works on four samples simultaneously, which are at different stages of completion. Uniform sample treatment and rapid data collection, with minimal human intervention, were achieved through serial sample preparation and analysis.

Accuracy and precision data resulting from this new PySystem were similar to those obtained with the manual methods, as well as those obtained from earlier custom systems. Some major advantages of the PyTechnology system over custom systems are: quicker start-up and implementation due to less end-user programming required for system installation, faster turn-around times from one method or application to another because of the Pysystem architecture, and better system reliability and less down-time due to more built-in error recovery routines.

Application

The applicability of the automated methodology/technology was demonstrated by analyzing plasma samples from various clinical studies with the new automated methods. Resulting plasma concentration versus time profiles obtained from normal volunteers and/or patients receiving doses of the various compounds are shown in Figures 8-11.

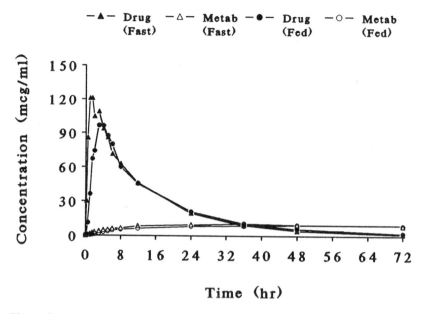

Figure 8. Plasma level profiles from a normal volunteer after receiving a 900-mg oral dose of the drug (prinomide) following a 10-hour fast and a substantial meal using the PyTechnology robotic system for sample analysis.

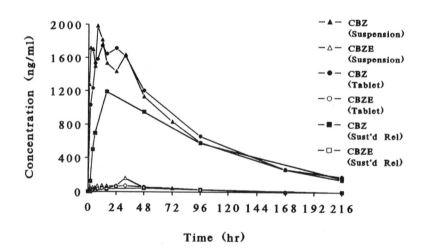

Figure 9. Plasma level profiles from a normal volunteer after receiving single oral doses of several formulations of 200-mg carbamazepine using the PyTechnology robotic system for sample analysis.

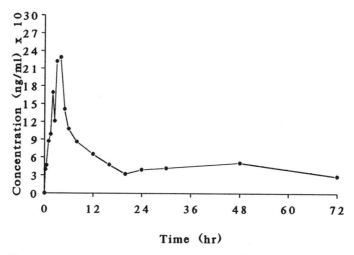

Figure 10. Plasma level profile from a normal volunteer after receiving an experimental controlled-release formulation dose of 150-mg diclofenac sodium using the Zymate II robotic system for sample analysis.

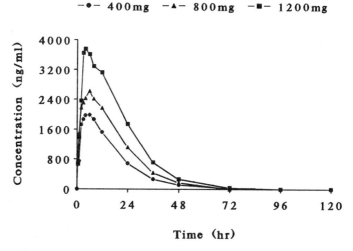

Figure 11. Mean plasma level profiles from normal volunteers (N=3) after receiving single oral doses of 400-1200 mg of the drug, CGP 33101 using the Zymate II robotic system for sample analysis.

Conclusions

Automated analytical methodology employing HPLC for sample analysis has been developed and validated for quantifying various anti-inflammatory and anti-epileptic compounds, as well as their respective metabolites, in human plasma samples originating from drug disposition studies. The methods have been successfully applied to the analysis of plasma samples from healthy volunteers and/or patients participating in ongoing clinical trials involving the indicated compounds.

Approximately 1000 analyses, including standard curve samples, quality control samples, clinical samples and re-analyses can be completed in about two months of continuous operation (four days per week) of the systems.

Literature Cited

1. Brunner, L.A.; Luders, R.C. In *Advances in Laboratory Automation - Robotics*; Strimaitis, J.R.; Hawk, G.L., Eds.; Zymark: Hopkinton, MA, **1988**; Vol. 5, pp. 413-31.
2. Luders, R.C.; Brunner, L.A. *J. Chrom. Sci.* **1987**, 25(5), pp. 192-7.
3. Brunner, L.A.; Luders, R.C. *J. Chrom. Sci.* **1991**, 29(7), pp. 287-91.

RECEIVED February 24, 1992

Chapter 14

Preparative Separations of Glycerophospholipids by High-Performance Liquid Chromatography

J. V. Amari and P. R. Brown

Department of Chemistry, University of Rhode Island, Kingston, RI 02881

Preparative high performance liquid chromatography (HPLC) is generally the method of choice to isolate and purify large quantitites of naturally occurring or synthetic glycerophospholipids. Separations by class or molecular species can be obtained with high purity yields. Separations by class are usually accomplished with gradient elution but separations by molecular species frequently require multi-chromatographic steps. In this article various chromatographic techniques for separations of phospholipids on the preparative scale will be reviewed.

In the pharmaceutical industry, glycerophospholipids (phospholipids) are used as natural emulsifiers, wetting agents and diuretics. In addition they are precursors in the synthesis of novel pharmaceuticals and they aid in delivery of drugs. Although they are mainly found in cell membranes, they are also present in tissues, body fluids and organs. They influence many metabolic and enzymatic reactions and regulate transport systems. In addition membrane fluidity is highly dependent upon not only the class of phospholipids but also the species within each class.

(R)-Glycerophospholipids are derived from glycerol. They contain two acyl fatty acid chains (R_1 and R_2) of varying lengths and degrees of unsaturation, esterified to the first (sn-1 position) and second (sn-2-position) hydroxy groups of glycerol. The third hydroxy group forms an ester bond with phosphoric acid to give the structurally specific glycerophospholipid, phosphatidic acid (PA). In the large majority of phospholipids an

0097–6156/92/0512–0173$06.00/0

alcohol group is esterified to the phosphorous acid; for example, phosphatidylcholine (PC), phosphatidylethanolamine (PE) and phosphatidylinositol (PI). Structures of the most common phospholipids are shown in Figures 1-3.

Liponucleotides are a class of phospholipids which contain a nucleoside instead of an alcohol esterified to the phosphoric acid (Fig. 4). The naturally occurring liponucleotides, cytidine diphosphate diglyceride (CDP-DG) and its deoxycytidine analog (CDP-DG) have essential roles as intermediates in the biosynthesis of phospholipids(1-3). Synthetic liponucleotides(4-8) are of interest to the pharmaceutical industry as anti-tumor and anti-viral agents.

Because of the importance of phospholipids, good preparative scale separation methods are required in industry to isolate and purify these compounds by class and by individual molecular species. In analytical work thin layer chromatography (TLC) and gas chromatography (GC) have been used extensively to separate these compounds. These techniques are not suitable for preparative work. However, with high performance liquid chromatography (HPLC) large quantities of phospholipids can be separated rapidly with minimal degradation.

Preparative HPLC can be defined as any HPLC separation in which fractions of the effluent are collected. The collected solutes are subsequently used as reference standards, as products, or in the biosynthesis of other products. Columns range in size from the traditional analytical dimensions (15-30 cm x 0.20-0.46 cm i.d.) to preparative scale (15-200 cm x > 15 cm i.d.), and milligrams to multi-kilograms can be isolated. For small scale preparative separations, conventional analytical instrumentation is used. As the column size is increased, special HPLC equipment becomes necessary. The theory and applications of preparative HPLC are discussed in several texts(9-11).

Class Separations

Naturally occurring glycerophospholipids do not exist as a single discrete compound but are composed of multimolecular species of varying fatty acid chain lengths and degrees of unsaturation. Since the separation of phospholipids by class is based upon the polar head, silica has been the stationary phase most commonly used both in preparative and analytical HPLC.

The mode of chromatography associated with silica packings is adsorption chromatography. In the classical sense, adsorption chromatography, also called normal phase chromatography, is carried out with non-polar eluents which can be modified with more-polar solvents, such as

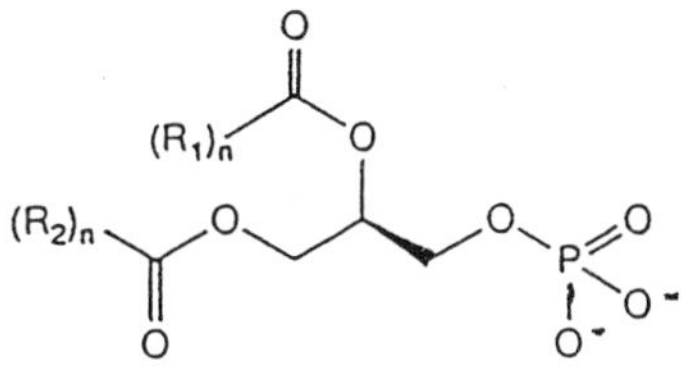

Phosphatidic Acid (PA)

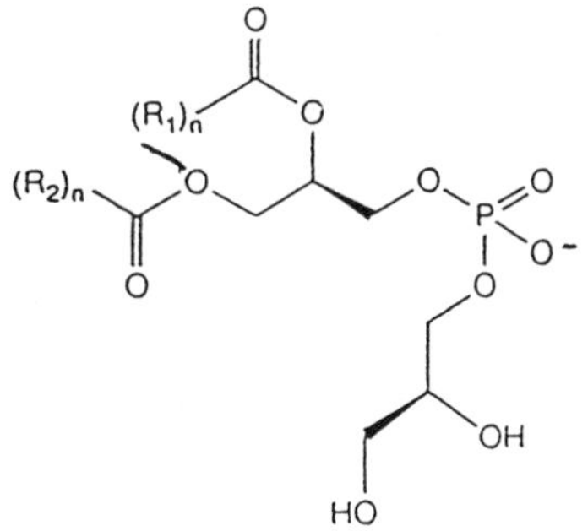

Phosphatidylglycerol (PG)

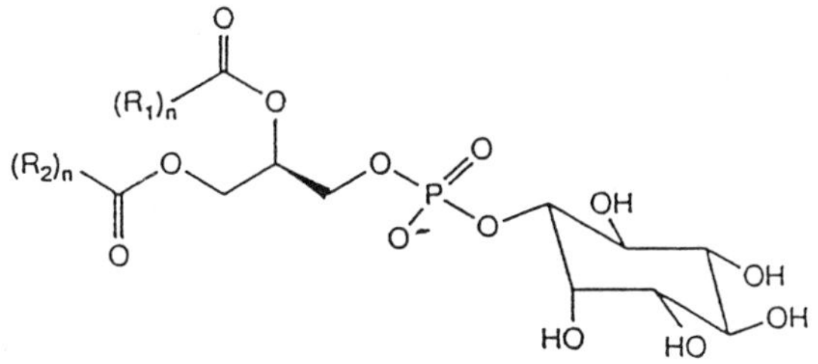

Phosphatidylinositol (PI)

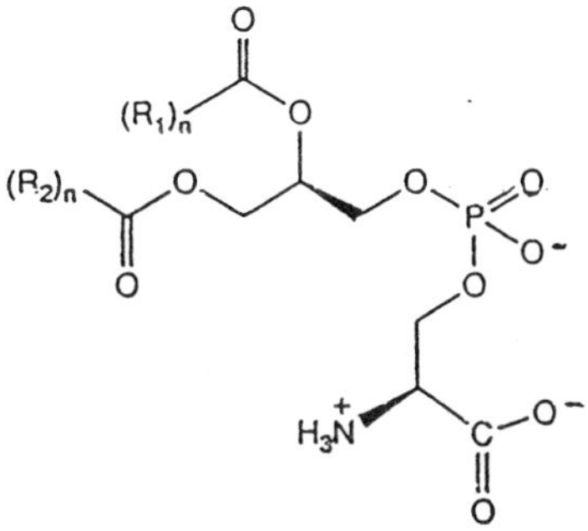

Phosphatidylserine (PS)

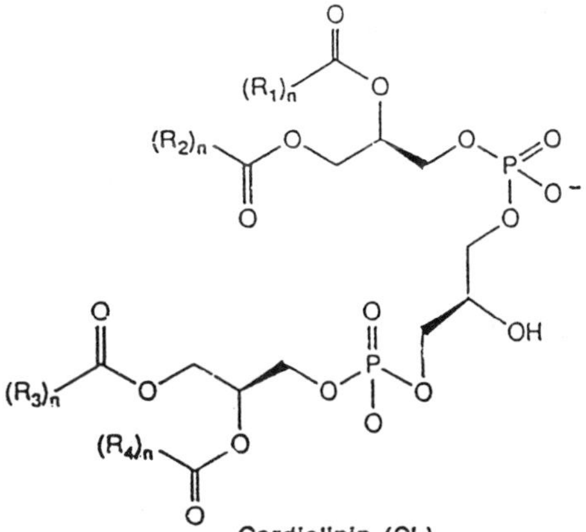

Cardiolipin (CL)

Figure 1.   Glycerophospholipid Structures. The R groups are
            acyl fatty acid chains of varying lengths and
            degress of unsaturation.

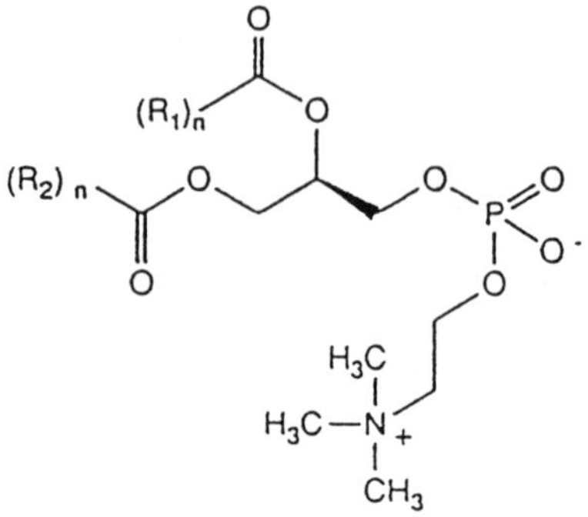

Phosphatidylcholine (PC)

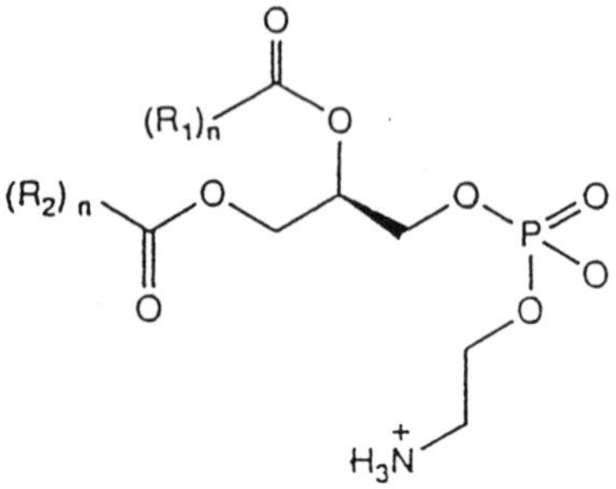

Phosphatidylethanolamine (PE)

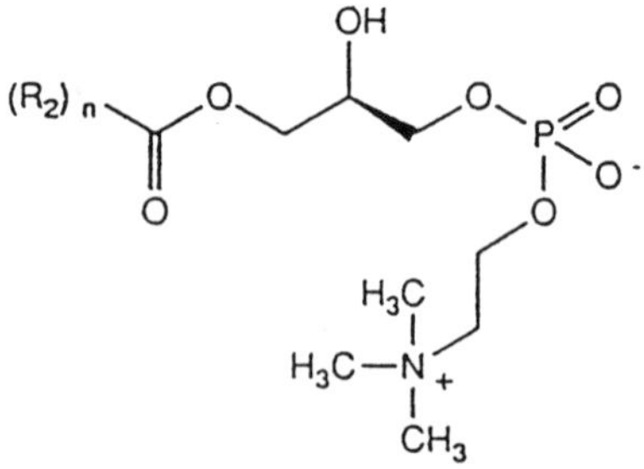

Lysophosphatidylcholine

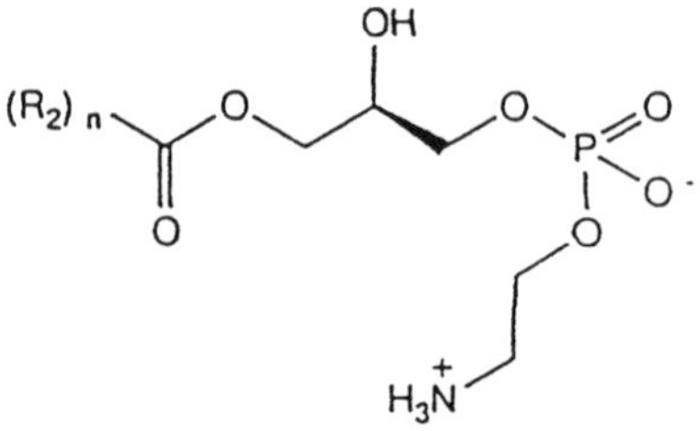

Lysophosphatidylethanolamine

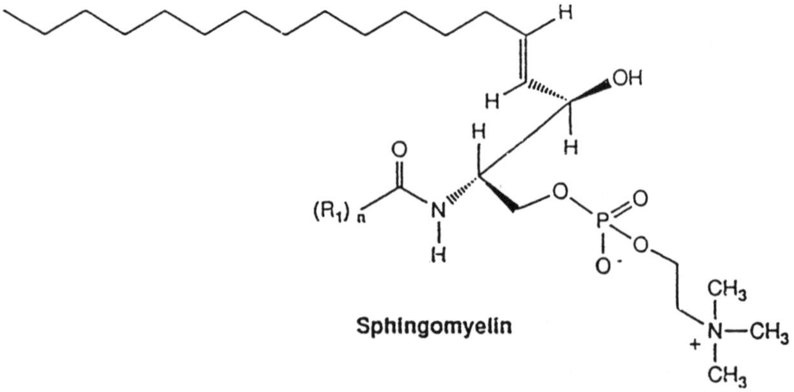

Sphingomyelin

Figure 2.   Glycerophospholipid Structures (continued)

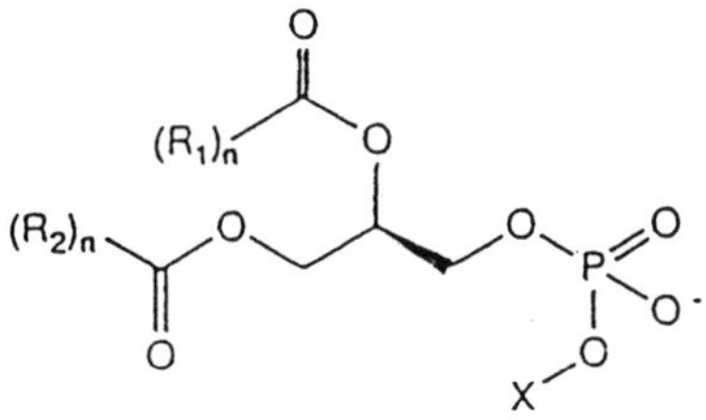

Diacyl

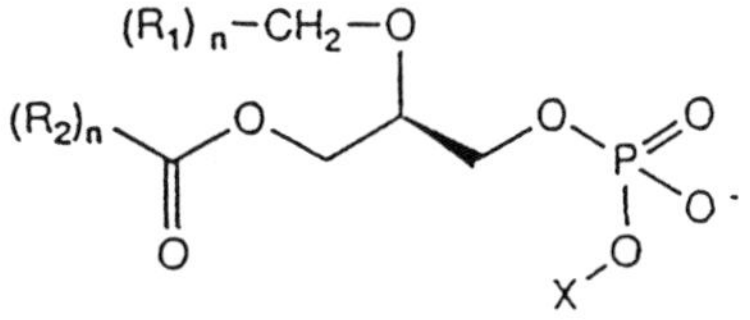

Alkylacyl

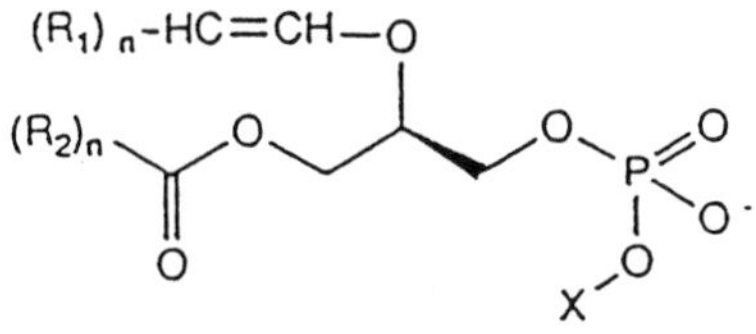

Alkenylacyl

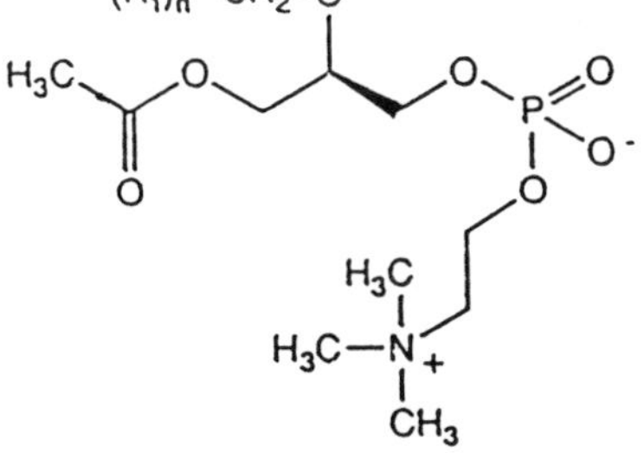

Platelet Activating Factor (PAF)

Figure 3.   Glycerophospholipid Structures (continued)

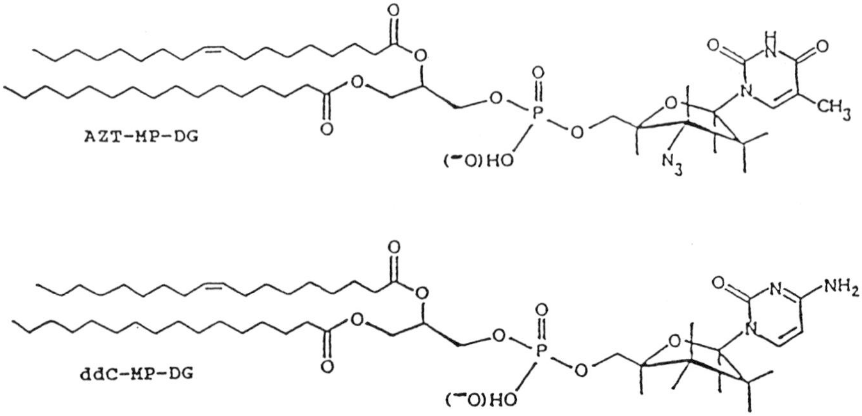

Figure 4.   Liponucleotide Structures

chlorinated solvents and/or alcohols. The silica packing is polar in nature and retention is based upon interaction of the polar moiety(s) of the solutes with the stationary phase(9,10). These interactions are complex(9,12,13). Since the non-polar diglyceride chains do not possess any significant attraction to the polar stationary phase and the attraction is very strong between the head groups and the silica, non-polar mobile phases have to be modified with polar solvents to decrease k' (capacity factor) values. Large k' values increase analysis time, collection time and band broadening, thus reducing sensitivity. Although reversed phase packings have been used extensively in analytical separations of phospholipids, only recently have they been used in preparative work(14-17).

In 1977, Fager et al.(18) achieved a large scale purification of PE, LPE and PC with 20-40 μm silica packed into eight 1 meter x 1 cm i.d. columns connected in series. The phospholipid were derived from egg yolks and 10 g samples were injected. The classes were eluted with a flow rate of 5 mL/min and increasing concentration of methanol (MeOH) in a CHCl₃-MeOH gradient. A total of 844 fractions were collected at three minute intervals, the final mobile phase was composed of 50% MeOH. Each fraction was spotted on a TLC plate and the purity determined by phosphate and amino acid analysis. Based upon the TLC analysis, the PE was eluted with baseline resolution at 25% MeOH, LPE at 40% MeOH and PC at 50% MeOH. Using a rapid isocratic system, Patel and Sparrow(19) purified large quantities of PC and PE with radially compressed silica columns (30 cm x 5.7 cm i.d.). The mobile phases consisted of a solvent system of CHCl₃ MeOH and H₂O; for PC the system was 60:30:4 CHCl₃-MeOH-H₂O by volume and for PE it was 60:30:2. A refractive index detector was used to monitor the separation because the CHCl₃ in the mobile phase prohibited the use of a UV detector. From twenty-six grams of the crude egg phospholipid, 10 g of PC was recovered; from a mixture of the neutral lipids and PE, 3.8 g of PE were recovered. Each separation was accomplished in less than 20 minutes. The column had to be flushed thoroughly to remove any highly retained components in the egg yolk.

To isolate and purify PC from egg yolk, Guerts Van Kessel et al.(20) used a 50 cm x 5 cm i.d. column packed with silica gel and the water was eliminated from the mobile phase. In this isocratic separation, a 60:40 solution of (CHCl₃:CH₃OH) was used with a recovery of 4.5 g. The PC was rechromatographed resulting in a final recovery of 4.3 g. The same system was used with a 70:30 mixture to isolate and purify synthetic PC and PE. To obtain purified products, the PC required rechromatography but the PE did not have to be rechromatographed. From 3.8

g of the crude mixture, 3.12 g of PC and 1.10 g of PE were recovered. Although relatively high recoveries were achieved, long separation times (3-4 hours) were necessary.

Smaal etal.(21) used preparative HPLC on a 25 cm x 2.2 cm i.d. column as a final purification step for beef heart CL after precipitation and column chromatography procedures. An eluent consisting of isopropanol (IPA), cyclohexane and H_2O (45:50:5) with RI detection was used. From 1 kg beef heart, 1.5-2.1 g of CL, which had a purity of 99%, were obtained.

With UV opaque solvents, refractive index (RI) detectors must be used. However this detector lacks the sensitivity necessary to monitor minor components or trace impurities. Therefore, analytical separations, which had been developed with UV transparent mobile phases(22-24) were adapted to preparative separations. Unfortunately, phospholipids are soluble in only a limited number of UV transparent solvents. In 1986 Hurst et al.[25] used gradient elution for isolating PI, PE and PC from crude soybean extracts. The mobile phases consisted of a constant amount of phosphoric acid (1.2%) and varying proportions of acetonitrile (ACN) and MeOH. The solvents system permitted detection at 205 nm. The maximum load was 118 mg on a 30 cm x 5.7 cm i.d. column packed with 55-105 μm silica. This mobile phase has been modified by adding other solvents or salts. For example Bahrami et al.(26) added IPA, H_2O and trifluoacetic acid (TFA). With an eluent of ACN-IPA-MeOH-H_2O-TFA (1.35:20:10:67:0.85) and a column packed with 55-105 μm silica, PC from rat lung tissue and lung lavage was isolated and the purity was determined by an assay for inorganic phosphate and GC analysis. Recently, another isocratic mobile phase consisting of 5 mM ammonium acetate (NH₄OAc) in ACN-IPA-MeOH-H_2O (80:13:5:12) was used to isolate PC from chicken egg yolk[27]. The NH₄OAc sharpened the peaks considerably, thereby increasing the load that could be injected per run. In addition, a user packed preparative Annular Expansion (A/E™) column (20 cm x 5.0 cm i.d.) was compared with a high-pressure packed, semi-preparative column (20 cm x 1.93 cm i.d.). Each column was packed with 15-30 μm silica. The maximum load of crude phospholipid on the high- pressure packed, semi-preparative column was 35 mg, with 25 mg of purified PC (99%) recovered. The load injected and flow rates were directly scaled up to the A/E™ column; 240 mg of crude phospholipids were injected and 130 mg of purified PC (99%) was recovered. The separations were virtually identical for each column and highly purified PC was obtained.

Guerts Van Kessel etal.(23) developed an analytical separation with UV transparent solvents which contained hexane (Hex). The mobile phase of 6:8:2 Hex-IPA-H_2O was then scaled up to purify SPH from beef erythrocytes isocratically(24). A 5 μm, 25 x 0.9 cm i.d. column was used and a sample of 10 mg was injected. The SPH was split into three peaks. Fatty acid analysis revealed that the longer chains were eluted in the first peak, while the shorter chains were more retained. This solvent system was applied by Ellington and Zimmerman(28) to isolate gram quantities of microsomal glycerophospholipids for preparing model membranes. In the linear gradient, the water was increased while the hexane and isopropanol were kept constant. The phospholipids PE, PI, PS, PC and SPH were well resolved (Figure 5) and up to 200 mg could be injected. A similar gradient was used to fractionate algal derived phospholipids(14); however, in the initial solvent only Hex and IPa were present (6:8) and water was added throughout the gradient program. The phospholipids were baseline resolved with a load of 17.5 mg.

In addition to solvent programing, flow programing can increase the column load(29). A study on two types of silica was conducted to determine the influence of water on the load ability, retention time and peak shape. The water content in the mobile phase plays an important role in the degree of retention for phospholipids. Very high k' values were observed for PE, PC and PI if the water content was less than 5%. On the other hand for PE the k' values were acceptable with less than 5% water. Therefore a step gradient was incorporated, with an Hex-IPA-H_2O (55:44:4) eluent which allowed the early eluting phospholipids to be resolved. Then the solvent was changed to a ratio of 55:44:5.7. The step gradient was used with timed increases in flow-rate. Only one type of silica gave adequate scaled up separations with the conditions developed and a 100 mg load.

Juaneda and Roquelin(30) found that for complete separation in preparative isolations, two HPLC methods were required to obtain milligram quantities of CE, PC, PE, PI, PS, LPC, and SPH for additional analysis and membrane models. The phospholipids in human heart were separated with a IPA-Hex-H_2O gradient (54:41:5 to 52:32:9). All the phospholipids were well separated except PC and SPH, which were coeluted. The PC/SPH fraction was resolved by rechromatography with an isocratic mixture of ACN-MeOH-H_2O (71:21:8). The separations were completed within 90 minutes.

Milligram quantities (1-2 mg) of PE, PI, PS and PC were isolated on an analytical column (25 cm x 0.46 cm i.d.) by gradient elution with a combination of the mobile phases by Jungalwala et al.(22) and Guerts Van Kessel etal.(23). Solvent A consisted of Hex-IPA-ACN-H_2O (364:486:94:56) and solvent B, in which there was no acetonitrile, had a composition of Hex-IPA-H_2O

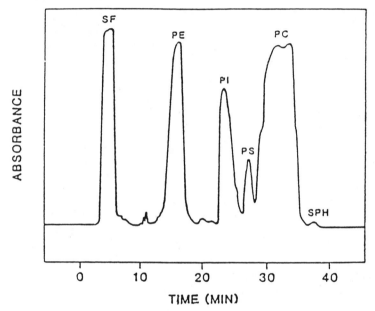

Figure 5.   Elution of extracted microsomal phospholipid by
            preparative HPLC.   Reproduced with permission
            from reference 28, copyright 1987.   John S.
            Ellington.

(394:526:80). This method was adapted to isolate 1-2 mg of phospholipids from cod tissue.[31] With a 25 cm x 0.46 cm i.d. column and an initial mobile phase of solvent A, PE and PI were eluted. After switching to solvent B, PS and PC were eluted.

Glass(15,16) replaced hexane with isooctane for the isolation of PC derived from soy beans. A semi-preparative isolation of 1.25 mg of PC was achieved in 20 minutes with an isocratic eluent of isooctane-IPA-H_2O (40:53:7) at a flow rate of 4 mL/min[15]. The separations were accomplished on a 15 cm x 1.0 cm i.d. column packed with 5 μm silica. Glass[16] further modified the mobile phase by a reduction of isopropanol and an increase of water (40:51:9) along with a corresponding decrease in flow rate. The optimum load was 5 mg with a 15 cm x 1.0 cm i.d. column. With a 2% increase in water, a 2% decrease in isopropanol and a decrease in flow rate to 2 mL/min, the k' of PC was similar to that obtained with the original mobile phase composition (40:53:7) at 4 mL/min.

A rapid isocratic separation was developed to purify PAF and Lyso-PAF from human skin(32). A diol phase was used in a 25 cm x 0.46 cm i.d. column with a buthylmetylether-MeOH-H_2O-NH_3 (200:100:10:0.02) mobile phase. From 1 g of human skin 20-200 ng of PAF-like material was recovered. Andrikopoulos et al.(33) reported that up to 2.5 mg of semisynthetic PAF and lyso-PAF could be separated per chromatographic run using a mobile phase of ACN-MeOH-H_2O (300:150:35) and flow programing to reduce the separation time.

Species Separations

Within each glycerophospholipid class a number of molecular species exists due to the varying lengths and degrees of unsaturation of the fatty acid chains. The separation of multi-molecular species differing in chain length or degrees of unsaturation is a formidable task, because as many as thirty or more species may be present. With many similar components present, highly efficient packings are necessary and particle sizes of less than 12 μm are routinely used. Species separations are based upon the hydrophobicity and degree of unsaturation of the fatty acid chains with reversed-phase and/or argentation chromatography. Reversed-phase supports, composed of C_{18} stationary phases, have been used exclusively to resolve individual molecular species on a preparative scale(14-17), although there has been a report of partial species separation by silica(18). The resolution of the molecular species on C_{18} supports are based primarily upon the hydrophobic interactions of the diglyceride chains with the alkyl ligand of the stationary phase. Predictions may be made on retention and the elution order of species.

The retention progressively increases as the number of carbon atoms in the homologous series increases; however the retention decreases as the number of double bonds in the same fatty acid chains increases. Unfortunately every molecular species cannot be resolved on C_{18} columns. With species separation, not only is the separation difficult but sensitive detection is required. At present, UV detectors are the most sensitive, stable, rugged and easily used for preparative species separation. The UV response is primarily due to the double bonds in the fatty acid chains; this response is very low for saturated chains. In order to use UV detection, the mobile phase must be transparent at the working wavelengths (200-210 nm). This requirement and the restraints caused by the solubility of the phospholipids greatly limits the choice of mobile phases.

Only a partial species resolution of PE, PLE and PC was reported on silica(18). Although discrete peaks were not observed, the composition of the species were determined by GC analysis of the collected fractions. Separation profiles were constructed based on phospholipid concentrations versus fraction number. The profile revealed partially resolved peaks within the PE and PC peaks. High concentrations of polyunsaturated acyl side chains were present in the early fractions of PE, LPE and PC, while later fractions consisted of mono- and di-saturated as well as saturated acyl chains.

Individual PC species were required for calibration factors in the quantitation by HPLC of human bile PC species(17). Egg yolk PC (15 mg) was fractionated into 15 distinguishable peaks on a 30 cm x 4 cm i.d. column packed with 7 μm C_{18}. Solvent A consisted of 100% MeOH and B was 20 mmol/L choline chloride in MeOH-H_2O-ACN (90:8:3); thus UV detection at 205 nm could be used. Each peak was composed of more than one molecular species, based upon the fatty acid composition determined by GC and HPLC analysis. Twenty-three probable PC molecular species were identified. Rezonka and Podojil(14) separated species of the individual glycerophospholipids (PE, PG, PI and PA) and PC derived from alga Chlorella kessleri on a 5 μm, 25 cm x 2.12 cm i.d. column. Choline chloride (20 mM) was used to modify the MeOH-H_2O-ACN (90.5:7:2.5) solvent system. The maximum load for each glycerophospholipid was 5 mg in 100 μL. Eighteen molecular species were well resolved for PC. As determined by GC-MS, the distribution of molecular species for each class varied.

Regioisomers of LPC were isocratically separated with a mobile phase comprised of 20 mM choline chloride in MeOH-H_2O-ACN (57:23:20)(34). It was possible to separate 1-acyl-sn-glycerol-3-phosphocholine from 2-acyl-sn-glycerol-3-phosphocholine. The k' values were lower for LPC isomers when the acyl chain was in the sn-2 position rather than the sn-1 position. Utilizing a 25 cm x 2 cm i.d. column, 20-30 mg of the isomeric LPC species were isolated. Under similar chromatographic conditions, bovine heart and brain LPE and soy bean LPC were separated into molecular species. Milligram quantities of starting material were provided for subsequent use in the synthesis of homogeneous diracylphospholipid probes[35]. Up to 2 mg of the lysoglycerophospholipids could be injected without overloading the column.

Ammonium acetate has been previously defined has also been successfully incorporated into the mobile phase(16,36); typically, 100 mM of NH_4OAc was used. The pH, although undefined in organic solvents, was reported to be 7.4 in the aqueous constituent. A semi-preparative separation of PC species from soy beans was reported with a mobile phase of 100 mM NH_4OAc in $MeOH-H_2O$ (95:5)[16]. With 2.1 mg injected, five peaks were observed in the chromatogram (Figure 6). The highly unsaturated species, eluted in peak 1, while the more saturated species eluted in peak 5. Polyunsaturated glycerophospholipid species in PC and PE were isolated along with the corresponding oxidation products with a 25 cm x 2.25 cm i.d. column[36]. Up to 20 mg were purified isocratically with 200 mM NH_4OAc in 100% MeOH.

Liponucleotides have recently been synthesized as experimental anti-AIDS drugs. The most promising of these compounds are AZT-MP-DG and ddC-MP-DG, which have been synthesized, and work is in progress on the synthesis of ddI-MP-DG. Preparative HPLC was used to isolate and purify enough of these compounds for characterization by spectroscopic methods and to test for antiviral activity. In addition, large amounts of the intermediate, AZT-P, was needed in the synthesis of AZT-MP-DG[37]. To isolate both the intermediate and final product the preparative separations were achieved with reversed-phase supports(38). A column 25 cm x 1 cm i.d., packed with 10 μm C_{18}, was used and the eluent consisted of MeOH-1mM KH_2PO_4, pH 2.4, (93:7) for AZT-MP-DG and 95:5 for ddC-MP-DG. Purities of greater than 99% were obtained when the load was not greater than 25 mg of the crude synthetic mixture.

Conclusions

Glycerophospholipids are a complex class of lipids. Because of their importance in the pharmaceutical industry, isolation and purification of large quantities of various types of phospholipids are required. In order to resolve phospholipids by class, many stationary phases are available and gradient elution is usually necessary. Resolution of molecular species is more complex than the separation by class. For high purity separation of

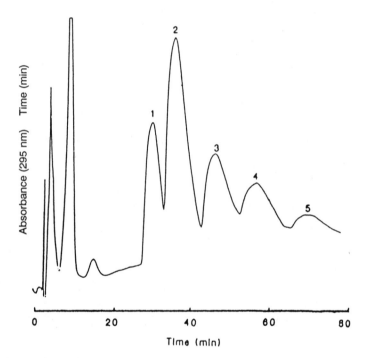

Figure 6.   Chromatogram of the soybean PC into its molecular
             species   on a   semipreparative   ODS   column.
             Reproduced with permission from reference 16,
             copyright 1991, Marcel Dekker, Inc.

molecular species, multi-chromatographic steps are usually required. The type of solvents and modifiers in the mobile phase is often limited by solubility constraints and the type of detector used. Although multiple separation and detection techniques may be necessary, preparative HPLC is the technique of choice to isolate and purify large quantities of naturally occurring or synthetic phospholipids either by class or molecular species.

REFERENCES

1) Raetz, C.R.H. and Kennedy, E.P. Journal of Biological Chemistry, 248 (1973) 1098-1105.
2) Hauser, G. and Eichberg, J. Journal of Biological Chemistry, 250 (1975) 105-112.
3) Thompson, W. and Macdonald, G. Journal of Biological Chemistry, 250 (1975) 6779-6785.
4) Raetz, C.R.H., Chu, M.Y., Srivastava, S.P. and Turcotte, J.G. Science, 196 (1977) 303-305.
5) Turcotte, J.G., Srivastava, S.P., Meresak, W.A., Riztralla, B.A., Louzon, F. and Wunz, T.P. Biochimica et Biophysica Acta, 619 (1980) 604-618.
6) Turcotte, J.G., Srivastava, S.P., Steim, J.M., Calabresi, P., Tibbetts, L.M. and Chu, M.Y. Biochimica et Biophysica Acta, 619 (1980) 619-631.
7) Steim, J.M., Neto, C.C., Sarin, P.S., Sun, D.K., Sehgal, R.K. and Turcotte, J.G. Biochemical and Biophysical Research Communications, 171 (1990).
8) Hostetler, K.Y., Stahmiller, L.M., Lenting, H.B.M., Vandem Bosch, H. and Rideman, D.D. Journal of Biological Chemistry, 265 (1990) 6112-6117.
9) Snyder, L.R. and Kirkland, J.J. Introduction to Modern Liquid Chromatography 2nd ed. John Wiley & Sons, Inc., New York: 1919 CH 15.
10) Bidlingmeyer, B.A. in Preparative Liquid Chromatography Journal of Chromatography Library - Volume 38. Bidlingmeyer, B.A. ed. Elsevier, New York: 1987 CH. 1.
11) Colin, H. in High Performance Liquid Chromatography Volume 98 in Chemical Analysis. Brown, P.R. and Hartwick, R.A. ed. John Wiley & Sons, New York: 1989 CH 11.
12) Scott, R.P.W., in Advances in Chromatography. Giddings, J.C., Grushka, E., Cazes, J. and Brown, P.R. ed. Vol. 20 Marcel Dekker, Inc., New York: 1982.
13) Engelhardt, H., Muller, H. and Schon, U., in High Performance Liquid Chromatography in Biochemistry. Henschen, A., Hupe, K-P, Lottspeich, F. and Voelter, W. ed. VCH Publishers, Florida: 1985: CH 2
14) Rezonka, T. and Podojil, M. Journal of Chromatography, 464 (1989) 397-408.

15) Glass, R.L., Journal of Liquid Chromatography, 38
     (1990) 1684-1686.
16) Glass, R.L., Journal of Liquid Chromatography, 14(2)
     (1991) 339-349.
17) Cantafora, A., Dibaise, A., Alvaro, D., Angelico,
     M., Marin, M. and Attili, A.F., Clinica Chimica
     Acta, 134 (1983) 281-295.
18) Fager, R.S., Chapiro, S. and Litman, B.J., Journal
     of Lipid Research, 18 (1977) 704-709.
19) Patel, K.M. and Sparrow, J.T., Journal of
     Chromatography, 150 (1978) 542-547.
20) Guerts Van Kessel, W.S.M., Tieman, M. and Demel,
     R.A., Lipids, 16(1) (1981) 58-63.
21) Smaal, E.B., Romijn, D., Guerts Van Kessel, W.S.M.,
     de Kruijff, B. and de Gier, J., Journal of Lipid
     Research, 26 (1985) 634-637.
22) Jungalwala, F.B., Evans, J.E. and McCluer, R.H.,
     Biochemistry Journal, 155 (1976) 55-60.
23) Guerts Van Kessel, W.S.M., Hax, W.M.A., Demel, R.A.
     and De Gier, Jr., Biochimica et Biophysica Acta, 486
     (1977) 524-530.
24) Hax, W.M. and Guerts Van Kessel, W.S.M., Journal of
     Chromatography, 142 (1977) 735-741).
25) Hurst, W.J., Martin, Jr., R.A. and Sheeby, R.M.,
     Journal of Liquid Chromatography, 9(13) (1986) 2969-
     2976.
26) Bahrami, S., Gasser, H. and Redl, H., Journal of
     Lipid Research, 28 (1987) 596-598.
27) Amari, J.V., Brown, P.R., Grill, C.M. and Turcotte,
     J.G., Journal of Chromatography, 517 (1990) 219-228.
28) Ellington, J.S. and Zimmerman, R.L., Journal of
     Lipid Research, 28 (1987) 1016-1018.
29) Van de Meeren, P., Vanderdeelen, J., Huys, M. and
     Baert, L., Journal of the American Oil Chemists'
     Society, 67 (1990) 815-820.
30) Juaneda, P. and Rocquelin, R., Lipids, 21(3) (1986)
     239-240.
31) Lie, O. and Lambertsen, G., Journal of
     Chromatography, 565   (1991) 119-129.
32) Mallet, A.I., Cunningham, F.M. and Daniel, R.,
     Journal of Chromatography Biomedical Applications,
     309 (1984) 160-164.
33) Andrikopoulos, N.K., Demopoulos, C.A. and Siafaka-
     Kakpadai, A., Journal of Chromatography, 363 (1986)
     412-417.
34) Creer, M.H. and Gross, R.W., Lipids, 20 (1985) 922-
     928.
35) Creer,M.H. and Gross, R.W., Journal of
     Chromatography Biomedical Applications, 338 (1985)
     61-69.
36) Holte, L.L., van Kuijh, F.J. G.M. and Dratz, E.A.,
     Analytical Biochemistry, 188 (1990) 136-141.

37) J.G. Turcotte, P.E. Pivarnik, S.S. Shiral, H. K.
     Singh, R.J. Sehgal, D. MacBride, N-I Jang and P.R.
     Brown. Journal of Chromatography, 449 (1990) 55-
     61.
38) J.V. Amari, P.R. Brown, P.E. Pivarnik, R.K. Sehgal
     and J.G. Turcotte, Journal of Chromatography, 590(1)
     (1992) 153-161.

RECEIVED June 29, 1992

Chapter 15

Small-Molecule Pharmaceutical Separations by Capillary Electrophoresis
Method Development and Manipulation of Selectivity

Michael E. Swartz and John VanAntwerp

Applied Technology Group, Millipore Waters Chromatography, Milford, MA 01757

An approach for capillary electrophoresis (CE) method development is established and applied to small molecule pharmaceutical separations using analgesics as examples. The effect of various parameters on selectivity were evaluated, and some recommendations are made regarding successful quantitation. Parameters such as linearity, sensitivity, and reproducibility were also examined. This knowledge was extended to other application areas, and CE separations of such compounds as penicillins, steroids, water soluble vitamins, and enantiomeric compounds were developed. It is shown that the selectivity, sensitivity, and reproducibility of CE are adequate for routine use in the pharmaceutical laboratory.

Capillary Electrophoresis (CE) was first described by J.W. Jorgensen and K.D. Lukacs in the early 1980's (1-3). Since then, hundreds of articles have appeared in the literature describing the theory, instrumentation, and the applicability of the technique (4,5, and references therein). Much of what has been presented to date, however, has not been of much practical use, or indicative of how CE can be used as an analytical tool. High efficiencies and different selectivity make CE useful as an orthogonal technique which can complement other analytical methods. It therefore has a great potential as an analytical tool for complex pharmaceutical molecules and formulations.

Several parameters can be used to effect separations and manipulate selectivity in CE. These include the capillary dimension, buffer composition, ionic strength and pH, applied voltage, sample matrix, and buffer additives including organic solvents, surfactants, or ion-pairing reagents. When developing a method, each of these parameters should be investigated in a stepwise logical fashion. The method developed should not only provide the necessary selectivity and resolution, but quantitative reproducibility and sensitivity for all components in a reasonable analysis time.

In this work, the effects of buffer pH, ionic strength, applied voltage, and sodium dodecyl sulphate (SDS) concentration were systematically evaluated for their effects on migration time and selectivity in a separation of some common analgesic standards. The resulting conditions were then used to examine linearity, sensitivity, and reproducibility. This work was then extended to more complex samples to examine the applicability of CE to additional compound classes.

0097–6156/92/0512–0190$06.00/0

Experimental

Capillary Electrophoresis System. A Waters **Quanta 4000** Capillary Electrophoresis System was used throughout (Millipore Waters Chromatography, Milford, MA, USA). Separations were performed on capillaries of various dimensions (Waters) and are documented in the appropriate figure or table caption. All analyses were performed with UV detection at 214 nanometers, and hydrostatic injections (10 cm height). Data was collected on an 845 Chromatography Data Workstation (Millipore) at 10 points/second.

Chemicals And Reagents. Chemicals and standards were obtained from Aldrich Chemical Co. (Milwaukee, WI), and Sigma Chemical Co. (St. Louis, MO). All were obtained in the highest purity available and used without further purification. Background electrolytes were prepared with Milli-Q water (Millipore, Bedford, MA), and LC grade methanol (Baker, Phillipsburgh, NJ). Phosphate buffers were prepared using monobasic sodium phosphate and adjusted to the desired pH with sodium hydroxide. Phosphate/borate buffers were prepared using monobasic sodium phosphate and sodium tetraborate. Tris-phosphate buffers were prepared from Tris-HCl and adjusted to the desired pH with phosphoric acid. Samples identified in Figures as three letter codes are defined as folows: caffeine (CAF), acetominophen (AMP), acetylsalicylic acid (ACE), salicylic acid (SAL), and salicylamide (SAM).

Results and Discussion

Analgesic Separation Optimization, Free Zone. The separation of analgesic compounds by CE have been reported previously (6), however since the compounds of interest were somewhat different, and a more thorough understanding of the parameters governing the separation was desired, these conditions were used as a starting point for further method development. Figure 1 shows the effect of pH on the migration time for five common analgesic active ingredients. It can be seen that the best resolution of all five components occurs at pH's in excess of 10.0. Due to the proximity of a phosphate pK, a pH of 11.0 was chosen for subsequent optimization.

Although adequate resolution is obtained, there are two inadequacies with the separation obtained at this point in the method development. One is an exceedingly long analysis time. More important, however, is the fact that under these conditions, caffeine, and any other neutral compounds or excipients that would be present in a typical sample, will migrate with the electroosmotic flow. This does not allow for accurate quantitation of these compounds. Further optimization was therefore neccessary to reduce analysis time, and to selectivley migrate caffeine away from any other neutral species. One way to decrease analysis time is to increase the run voltage. This is illustrated in Figure 2. By choosing a pH of 11.0, and increasing the applied voltage, analysis time decreased significantly. All other conditions identical, increasing the applied voltage resulted in a 50% decrease in analysis time. Increasing the voltage also has the added advantage of increasing efficiency, however at the sacrifice of decreased resolution and generation of additional joule heating in the capillary. For the components of interest in this separation, resolution was not a limiting factor, and joule heating was adequately controlled through heat dissapation and judicious selection of buffer ionic strength. As presented in Figure 3, by decreasing the ionic strength at pH 11.0, analysis time again decreases, but at the sacrifice of resolution. For the purpose of this example, a voltage of 25 KV was chosen so that the entire range of ionic strength could be studied without experiencing problems with background current and joule heating. For the additional free zone data reported below, an ionic strength of 0.015 M and voltages of up to 30 KV were used as the best compromise between resolution and analysis time.

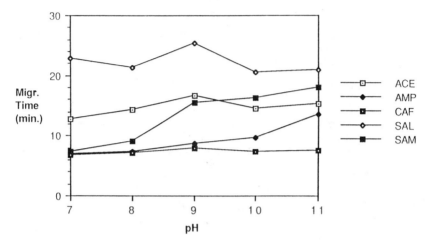

FIGURE 1: Effects of pH on migration time. A buffer of 0.05 M sodium phosphate at the specified pH and a 50 micron by 60 centimeter capillary operated at 15 KV was used. The samples were 0.1 mg/mL in water, with a 15 second injection time. Other conditions as outlined in the text (Methods and Materials). (Reproduced with permission from ref. 11. Copyright 1991 Marcel Dekker.)

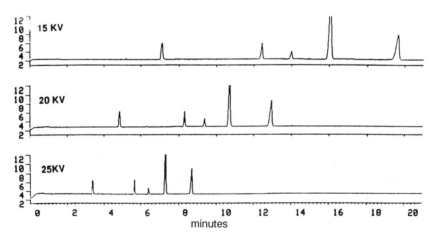

FIGURE 2: Effects of applied voltage on migration time. A pH of 11.0 was used, with voltage as specified; all other conditions are identical to those in Figure 1. (Reproduced with permission from ref. 11. Copyright 1991 Marcel Dekker.)

Linearity and Reproducibility, Free Zone. In order to be quantitative, a technique needs to be both linear and reproducible. By providing linearity over a wide dynamic range, the need for multiple level calibration curves is eliminated. In addition, criteria in use in the pharmaceutical industry requires linearity of up to three orders of magnitude. Using salicylamide as a representative sample, calibration curves were generated with serial dilutions from 1 mg/ml to 0.5 ug/mL that exhibited excellent linearity (Fig. 4).

Reproducibility was evaluated under similar conditions, and as reported in Table 1 the results obtained are well within requirements for accurate quantitation.

Table 1: CE Reproducibility

================ ======================================

75 micron by 60 cm Capillary

	Migr. Time	**Peak Area**	**Peak Height**
AVG	4.568	142073	34601
SDV	0.035	2770	714
%RSD	0.773	1.95	2.06

50 micron by 60 cm Capillary

	Migr. Time	**Peak Area**	**Peak Height**
AVG	5.894	31616	8361
SDV	0.051	504	193
%RSD	0.865	1.59	2.31

==

Conditions: Sample was a 0.1 mg/mL solution of Salicylamide, N = 9. A 15 second Hydrostatic injection was used, and a buffer of 0.02 M sodium phosphate, pH = 11.0,with an applied voltage of 25 KV. (Reproduced with permission from ref. 11. Copyright 1991 Marcel Dekker.)

Analgesic Separation Optimization, MECC. Selectivity is not affected by changes in voltage and ionic strength, however, and to successfully separate and quantitate neutral species it is necessary to investigate CE separation modes other than free zone. When sodium dodecylsulphate (SDS) is added to the buffer, a separation mode referred to as micellar electrokinetic capillary chromatography (MECC) dictates the separation mechanism (6). In MECC, nonionic species partition between the free solution and any micelles formed which are moving at different velocities in the capillary. Anionic species, on the other hand, are separated by the combination of electroosmotic flow and electrophoresis. The result is a separation of both ionic and nonionic species in the same run. In order to evaluate migration of nonionic compounds, a neutral marker such as methanol or formamide can be used. These compounds can be used as electroosmotic flow markers because they do not partition into the micelles, migrating strictly according to electroosmotic flow. The effect of changing the SDS concentration on the migration time of the analgesics is shown in Figure 5. The addition of SDS retards migration and also effects selectivity;

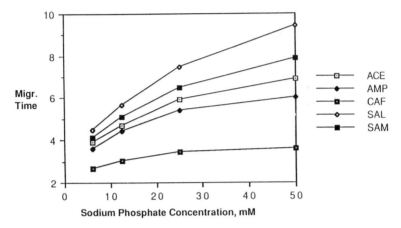

FIGURE 3: Effects of phosphate concentration on migration time. A pH of
11.0 and an applied voltage of 25 KV was used; all other conditions are identical
to those presented in Figure 1. (Reproduced with permission from ref. 11.
Copyright 1991 Marcel Dekker.)

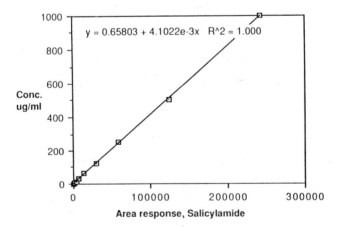

FIGURE 4: Linearity of response of CE detection. A buffer of 0.02 M sodium
phosphate pH 11.0, and a 50 micron by 60 cm capillary operated at 20 KV was
used. Sample is salicylamide, with an injection time of 10 seconds. (Reproduced
with permission from ref. 11. Copyright 1991 Marcel Dekker.)

salicylamide elutes fourth in free zone, but fifth in MECC as illustrated by the crossed lines in the graph. More significant, however, is that caffeine can now be separated from other neutral compounds as confirmed by separate injections of formamide used as an electrosmotic flow marker. An optimized separation that takes into account pH, voltage, ionic strength, and SDS concentration that can be used for quantitation within a realistic time frame can now be obtained (Fig. 6). When the ionic strength and SDS concentration are decreased at pH 11.0, higher voltages are possible resulting in an analysis time of under four minutes. It should be noted that an HPLC analysis of these same components, while providing completely different selectivity, is routinely performed by two separate methods, while CE accomplishes the separation of all components in a single analysis with significant time savings.

Sensitivity in MECC Mode. Detection limits for salicylamide were found to be 16 ng/mL (signal to noise ratio of three), obtained under the same conditions as those outlined in Figure 6. One commonly accepted requirement in the pharmaceutical industry is the ability to perform quantitatively at 0.1% levels with less than 10% RSD when evaluating impurity profiles. Figure 7 shows a separation of salicylamide where the scale has been expanded to show that impurities at the 0.1% level (peak area) can be detected. Table 2 summarizes the result of a separate quantitation of a salicylamide sample spiked with caffeine at the 0.1% level (w/w) that satisfies impurity profile specifications. To increase sensitivity it is possible, as in LC, to inject more sample. While linearity is still preserved, efficiency suffers significantly (Fig. 8,9). This compromise between sensitivity and efficiency could play a significant role in separation of complex multi-ingredient formulations or in the case of any closely resolved pair of peaks.

Table 2: Reproducibility of Trace Level Quantitation

===

	Area, Caffeine	Area, Salicylamide
	450	75969
	509	74628
	565	74673
	488	74274
	575	75113
	546	74441
	------	----------
AVG	522	74850
%RSD	9.3	0.8

===

Conditions: The buffer consisted of 0.015 M phosphate, pH 11.0, 0.025 M SDS. A 75 micron by 60 cm capillary operated at a voltage of 25 KV, with a hydrostatic injection time of 10 seconds was used. (Reproduced with permission from ref. 11. Copyright 1991 Marcel Dekker.)

Matrix Effects. Matrix effects play a more significant role in CE separations due to the fact that the capillary to injection volume ratio is lower than the equivalent ratio in other analytical separation techniques. Figure 10 illustrates the effect the sample matrix can have on the separation in the MECC mode. If the sample is prepared in 100% methanol (a typical extraction solvent), the localized micelle environment is disrupted, and the migration and peak shape of caffeine is drastically affected. This

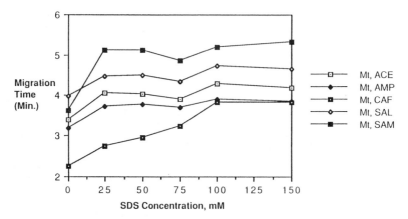

FIGURE 5: Effects of SDS concentration on migration time. A buffer of 0.02 M sodium phosphate, pH 11.0 (SDS concentration as specified), and a 50 micron by 60 cm capillary operated at 20 KV was used. The samples were 0.1 mg/mL in water, with a 15 second injection time. Other conditions as outlined in the text (Methods and Materials). (Reproduced with permission from ref. 11. Copyright 1991 Marcel Dekker.)

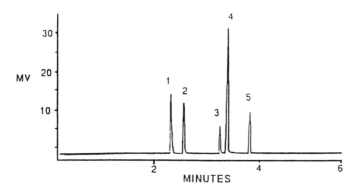

FIGURE 6: Optimized CE separation of the analgesics of interest. A buffer of 0.015 M sodium phosphate, pH 11.0, 0.025 M SDS, and a 50 micron by 60 cm capillary operated at 30 KVwas used. The samples were 0.2 mg/mL in water, with a 15 second injection time.   Peak identifications are: 1) caffeine, 2) acetaminophen, 3) acetylsalicylic acid, 4) salicylamide, and 5) salicylic acid. (Reproduced with permission from ref. 11. Copyright 1991 Marcel Dekker.)

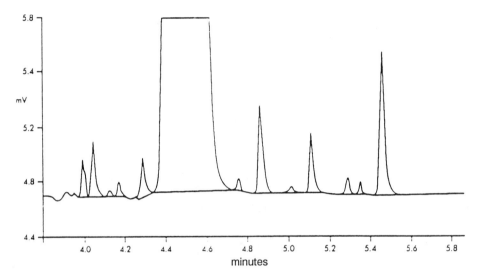

FIGURE 7: Salicylamide CE impurity profile. A buffer of 0.02 M sodium phosphate, pH 11.0, 0.075 M SDS and a 50 micron by 60 cm capillary operated at 20 KV was used. Sample is 0.1 mg/mL salicylamide in water, with an injection time of 10 seconds. Peaks are identified below as integrated from left to right. (Reproduced with permission from ref. 11. Copyright 1991 Marcel Dekker.)

Peak No.	Migr. Time	Area	Area %
1	3.99	377	0.13
2	4.04	515	0.18
3	4.12	29	0.01
4	4.16	140	0.05
5	4.23	390	0.14
6	4.52	283443	98.32
7	4.73	87	0.03
8	4.82	870	0.30
9	4.97	46	0.02
10	5.06	576	0.02
11	5.23	184	0.06
12	5.30	84	0.03
13	5.39	1532	0.53

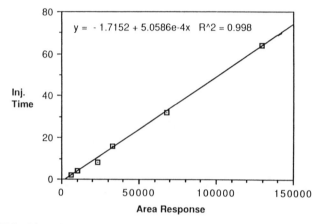

FIGURE 8: Linearity of injection time (volume) versus peak area. Conditions identical to those in Fig. 6. Injection time is in seconds. (Reproduced with permission from ref. 11. Copyright 1991 Marcel Dekker.)

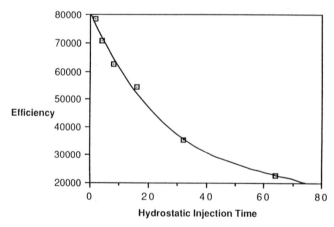

FIGURE 9: Effect of injection time (volume) on efficiency. Conditions identical to those in Figure 6. Injection time is in seconds. (Reproduced with permission from ref. 11. Copyright 1991 Marcel Dekker.)

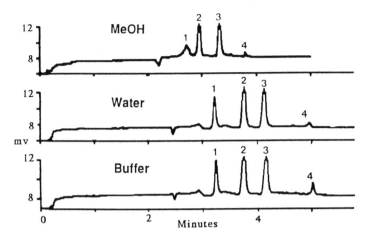

FIGURE 10: Effect of sample matrix effect on peak shape. A buffer of 0.015 M sodium phosphate 0.05 M SDS, and a 75 micron by 60 cm capillary operated at 18 KV was used. Sample was an acidic methanol extracted generic analgesic tablet (prepared by adding 10 mL/L 0.1 M HCl), with an injection time of 5 seconds. Peak identification is: 1) caffeine, 2) acetylsalicylic acid, 3) acetaminophen, and 4) salicylic acid (degradation product from peak 2). (Reproduced with permission from ref. 11. Copyright 1991 Marcel Dekker.)

would create problems with the reproducibility of integration, resulting in less accurate quantitative results. When the buffer and sample matrix are more evenly matched in composition, better peak shape and migration characteristics are obtained.

Additional Applications. Relying on the experience gained from the analgesic separation method development, and on recent literature reports for additional information (7-9), the separation of additional samples of different compound classes were attempted to further extend the applicability of this technique. Successful separations of penicillin antibiotics (Fig. 11) and some water soluble vitamins (Fig. 12), as well as steroids (Fig. 13) were obtained. The separation of the penicillins is of particular interest because it is possible to screen synthesis precursors, intermediates and all possible final products in a single run. The steroid separation uses a different additive, a bile salt, that is also capable of forming micelles. If this separation is performed in the prescence of SDS, all peaks co-migrate. For these applications, a combination phosphate/borate buffer was found to offer more buffer capacity for better reproducibility and peak shape for these particular compounds (*Swartz, unpublished results*).

Enantiomeric Separations. Another area of potential CE applicability is in the area of enantiomeric separations. Due to the inherent high efficiency of this technique, direct separations of enantiomers with buffer additives are possible by a number of different mechanisms (10). Figure 14 illustrates one such separation of (+/-)-ephedrine using a derivatized beta-cyclodextrin as a buffer additive to accomplish the chiral discrimination (*separation conditions adapted from reference 10*). The separation is performed under conditions of no electroosmotic flow, hence the migration of the compound of interest is governed strictly by electrophoretic migration of the charged analyte-cyclodextrin complex. It can be shown that pH and

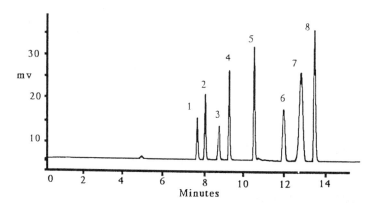

FIGURE 11: Separation of various penicillins by CE. A buffer of 0.02 M phosphate/borate pH 9.0, 0.05 M SDS, and a capillary 75 micron by 60 cm operated at 18 KV was used. The sample was 0.12 mg/mL each component in water, with a 5 second injection time. Peak identification is: 1) amoxicillin, 2) ampicillin, 3) 6-amino penicillanic acid, 4) oxacillin, 5) cloxicillin, 6) ticarcillin, 7) nafcillin, and 8) dicloxicillin. (Reproduced with permission from ref. 11. Copyright 1991 Marcel Dekker.)

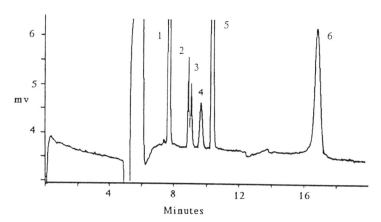

FIGURE 12: Separation of water soluble vitamins by CE. Conditions are identical to those in Figure 11. Sample is 0.1 mg/mL of each component in 50/50 water/methanol. Peak identification is: 1) vitamin B-6, 2) vitamin C, 3) pantothenic acid, 4) vitamin B-2, 5) niacin, and 6) vitamin B-1. (Reproduced with permission from ref. 11. Copyright 1991 Marcel Dekker.)

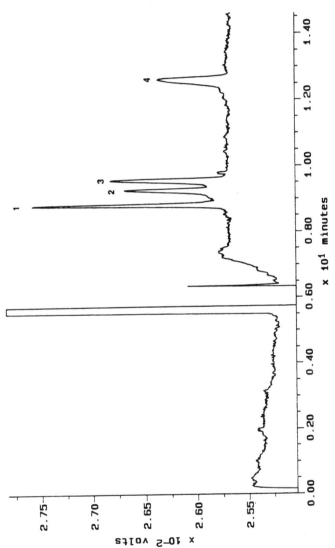

Figure 13: Separation of Corticosteroids by CE. A buffer of 0.02 M phosphate/borate pH 9.1, 0.1 M sodium cholate, and a capillary 50 micron by 60 cm operated at 20 KV was used. Sample is 0.25 mg/mL each component in methanol. Peak identification is: 1)triamcinolone, 2)hydrocortisone, 3) betamethasone, and 4) fluocinonide.

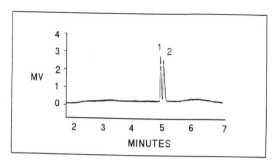

FIGURE 14: Separation of the enantiomers of ephedrine by CE. A buffer of 0.025 M Tris-phosphoric acid , pH 2.5, 0.015 M heptakis(di-O-methyl)-B-cyclodextrin/methanol, 80/20, and a capillary 50 micron by 35 cm operated at 18 KV was used. Sample was 0.1 mg/ML in water/methanol 50/50, with a 10 second injection time. Peaks 1 and 2 are (-) and (+) ephedrine respectively. (Reproduced with permission from ref. 11. Copyright 1991 Marcel Dekker.)

cyclodextrin type play an important role in the separation (*11*). The enantiomers of other chiral compounds including norephedrine, psuedoephedrine, phenylpropanolamine, methotrexate, and propranolol have been successfully separated under these and other similar conditions. (*Swartz, unpublished results*).

Conclusion

CE has been shown to have considerable potential for small molecule pharmaceutical separations. It has the required sensitivity, reproducibility, and selectivity to accomplish separations of widely diverse compound classes. While in most cases, CE provides complimentary information, it may be possible to dramatically decrease analysis times, and manipulate selectivity in ways that may provide information unobtainable by other techniques. It is possible, in an analogous fashion to HPLC, to systematically evaluate all of the parameters that can affect a separation in CE, and to arrive at a set of conditions that would provide rugged analyses for applications of this type.

References

1. Jorgenson, J.W., Lukacs, K.D., Anal. Chem., 53, 1298-1302 (1981).
2. Jorgenson, J.W., Lukacs, K.D., J. Chromatogr., 218, 209-216 (1981).
3. Jorgenson, J.W., Lukacs, K.D., Science, 222, 266-272 (1983).
4. Olechno, J.D., Tso, J.M.Y., Thayer, J., Wainright, A., Am. Lab., 22(17), 51-59 (1990).
5. Olechno, J.D., Tso, J.M.Y., Thayer, J., Wainright, A., Am. Lab., 22(18), 30-37 (1990).
6. Fujiwara, S., Honda, S., Anal. Chem., 59, 2773-2776 (1987).
7. Nishi, H., Tsumagari, N., Kakimoto, T., Terabe, S., J. Chromatogr., 477, 259-270 (1989).
8. Nishi, H., Tsumagari, N., Kakimoto, T., Terabe, S., J. Chromatogr. , 465, 331-343 (1989).
9. Nishi, H., Fukuyama, T., Matsuo, M., and Terabe, S., J Chromatogr., 513, 279-295 (1990).
10. Fanali, S., J. Chromatogr., 474, 441-446 (1989).
11. Swartz, M., E., J. Liquid Chrom., 14(5), 923-938 (1991).

RECEIVED May 18, 1992

INDEXES

Author Index

Affiliation Index

Subject Index

Production: Peggy D. Smith
Indexing: Deborah H. Steiner
Acquisition: Rhonda Bitterli
Cover design: Sue Schafer

Printed and bound by Maple Press, York, PA